## A Student's Guide to Phase Transitions

Phase transitions take place when a substance changes from one physical state to another, and they are of fundamental importance in science and engineering with applications ranging from superconductivity to climate science. This Student's Guide coherently examines the underlying dynamics of phase transitions, beginning with a detailed description of phase diagrams and their graphical interpretation, before introducing the van der Waals equations of state. It progresses to more advanced topics such as mean-field theory in magnetic systems, phase transitions in binary mixtures, and other more exotic types of phase transitions in liquid crystals, superconductors, and superfluids. A separate chapter covers the unique and subtle phase transition dynamics of water. The book includes numerous worked examples and problems, with full solutions available online. It will be a valuable resource for students and lifelong learners in the physical sciences and engineering.

JAMES S. WALKER is Emeritus Professor of Physics at Washington State University and was Boeing Distinguished Professor of Science and Mathematics Education at the time of his retirement. He has previously held positions at the University of Pennsylvania and the Massachusetts Institute of Technology. He is the author of a successful undergraduate textbook, *Physics* (Addison-Wesley), now in its fifth edition, and *A Student's Guide to the Ising Model* (Cambridge University Press).

**Other books in the Student Guide series:**

*A Student's Guide to the Laws of Thermodynamics*, Justin W. Garvin
*A Student's Guide to Rotational Motion*, Effrosyni Seitaridou, Alfred C. K. Farris
*A Student's Guide to the Ising Model*, James S. Walker
*A Student's Guide to the Navier–Stokes Equations*, Justin W. Garvin
*A Student's Guide to Special Relativity*, Norman Gray
*A Student's Guide to Laplace Transforms*, Daniel A. Fleisch
*A Student's Guide to Newton's Laws of Motion*, Sanjoy Mahajan
*A Student's Guide to the Schrödinger Equation*, Daniel A. Fleisch
*A Student's Guide to General Relativity*, Norman Gray
*A Student's Guide to Analytical Mechanics*, John L. Bohn
*A Student's Guide to Atomic Physics*, Mark Fox
*A Student's Guide to Infinite Series and Sequences*, Bernhard W. BachJr.
*A Student's Guide to Dimensional Analysis*, Don S. Lemons
*A Student's Guide to Numerical Methods*, Ian H. Hutchinson
*A Student's Guide to Waves*, Daniel A. Fleisch, Laura Kinnaman
*A Student's Guide to Lagrangians and Hamiltonians*, Patrick Hamill
*A Student's Guide to Entropy*, Don S. Lemons
*A Student's Guide to the Mathematics of Astronomy*, Daniel A. Fleisch, Julia Kregenow
*A Student's Guide to Vectors and Tensors*, Daniel A. Fleisch
*A Student's Guide to Data and Error Analysis*, Herman J. C. Berendsen
*A Student's Guide to Fourier Transforms*, J. F. James
*A Student's Guide to Maxwell's Equations*, Daniel A. Fleisch

# A Student's Guide to Phase Transitions

JAMES S. WALKER

*Washington State University*

CAMBRIDGE
UNIVERSITY PRESS

CAMBRIDGE
UNIVERSITY PRESS

Shaftesbury Road, Cambridge CB2 8EA, United Kingdom

One Liberty Plaza, 20th Floor, New York, NY 10006, USA

477 Williamstown Road, Port Melbourne, VIC 3207, Australia

314–321, 3rd Floor, Plot 3, Splendor Forum, Jasola District Centre, New Delhi – 110025, India

Cambridge University Press is part of Cambridge University Press & Assessment, a department of the University of Cambridge.

We share the University's mission to contribute to society through the pursuit of education, learning and research at the highest international levels of excellence.

www.cambridge.org
Information on this title: www.cambridge.org/highereducation/isbn/9781009664677

DOI: 10.1017/9781009664721



When citing this work, please include a reference to the DOI 10.1017/9781009664721

First published 2026

Cover image: Geysir geyser erupting in southwestern Iceland, Europe.
Smitt / iStock / Getty Images Plus.

*A catalogue record for this publication is available from the British Library*

*A Cataloging-in-Publication data record for this book is available from the Library of Congress*

ISBN 978-1-009-66467-7 Hardback
ISBN 978-1-009-66468-4 Paperback

Additional resources for this publication at www.cambridge.org/walker-phasetransitions

# Contents

# Preface

*A Student's Guide to Phase Transitions* is designed to explore the many ways that matter transforms from one physical state to another. These transformations – from solid to liquid, from liquid to gas, or from one magnetic state to another – govern everything from the weather in our atmosphere to the operation of high-tech devices.

This guide takes a structured, thermodynamics-first approach, recognizing that these transformations – referred to as *phase transitions* – are inherently about balancing energy and entropy. We begin by laying the groundwork with *phase diagrams*, the essential road maps that chart the conditions of temperature and pressure under which different phases are stable. You will learn to interpret these diagrams, including critical points, triple points, and boundary lines between the phases.

A major focus is the *Clausius–Clapeyron equation*, a powerful tool for calculating the slope of a phase coexistence curve and predicting how the boiling point or melting point changes with pressure. We then tackle the complexities of real gases and first-order phase transitions through the lens of the *van der Waals isotherm* and the *Maxwell construction*, showing how the theoretical instability predicted by the van der Waals equation is utilized to accurately depict the coexistence of liquid and gas phases.

No guide to phase transitions would be complete without a detailed look at the extraordinary and counterintuitive behavior of *water*. We explore why its liquid–solid phase boundary is uniquely tilted, and how its special properties – particularly its density maximum – are crucial to life on Earth.

The text also extends beyond single-component systems to explore the rich behavior of *binary mixtures*, introducing concepts such as azeotropes and eutectic points that are crucial for chemical engineering and materials science. We then pivot to the topic of *mean-field theory* as a simple yet powerful method for understanding magnetic phase transitions, such as the change from a paramagnet to a ferromagnet at the Curie temperature.

Finally, we broaden our scope to encompass a fascinating array of other phase transitions. This includes the intricate, temperature-dependent ordering found in *liquid crystals*, which form the basis of many modern displays, and the striking, zero-resistance phenomenon of *superconductivity*.

The goal of this book is to provide students with the conceptual tools and mathematical rigor to not only understand phase transitions, but to predict their behavior. Whether you are studying physics, chemistry, materials science, or engineering, the principles contained within this guide will prove invaluable. I am confident *A Student's Guide to Phase Transitions* will serve as a clear and engaging companion on your journey through the thermodynamics of phase transitions.

# Acknowledgments

I would like to thank my family, and especially my wife, Betsy, for their love and support. I would also like to thank the many students in my classes over the years, whose interest, enthusiasm, and questions have improved the presentation of the topics given in this book. Finally, I would like to thank the editorial staff at Cambridge University Press, and in particular Roxanne Daruwalla and Nicholas Gibbons, for their excellent work and patient guidance throughout the production of this book.

# 1

# Phases and Phase Transitions

Our lives are literally permeated by various phases of matter. Solid bones support our bodies, liquid blood flows through our veins, and the gaseous atmosphere is inhaled to infuse our lungs. In this chapter we explore these basic phases of matter, and the transitions that occur between them.

## 1.1 Phases

In condensed matter physics, the term *phase* has a specific meaning. It is not the phase angle that we talk about in electrical circuits, nor is it the phase associated with a quantum mechanical wave function. Instead, a phase can be thought of as a well-defined *state of matter*. Specifically, we define a phase of matter as follows:

> A *phase* is a uniform region of matter that has its own distinctive set of properties. Examples are *solids*, *liquids*, and *gases*.

Let's spell out some of the basic properties for the three phases just mentioned.

*Solid phase*: The most distinctive property of a solid is that it preserves its shape. The bones in our body maintain their shape, which allows them to support the weight of our bodies. Similarly, a sculpture carved from stone maintains the shape given to it by the sculptor. A solid's shape and size can be altered slightly when a force is applied to it, like stretching a spring, but it resists the deformation.

*Liquid phase*: A liquid's distinctive property is that it assumes the shape of its container, and flows readily from one place to another. For example, a liquid poured into a vessel fills it from the bottom up to the level determined by the amount of liquid present. Similarly, the liquid blood in our veins and arteries flows through a maze of blood vessels due to the pumping action of the heart. In the open air, liquids are splashy and support waves that can propagate across oceans. Liquids have a density similar to that of solids, and are practically incompressible.

*Gas phase*: A gas also assumes the shape of the container that holds it, but in contrast to a liquid, a gas fills the entire container from top to bottom. When we take a deep breath, the gases in the atmosphere fill and expand our lungs. Gases are low density and are readily compressed with an increase in pressure. They can expand rapidly when unconfined, like the air escaping from a popped balloon. As surprising as it may seem – having just ticked off the differences between liquids and gases – these phases can be considered as one and the same, referred to as the *fluid* phase. We will explore this in more detail later in this chapter.

Other phases also exist in nature. For example, a *plasma* is basically an electrically charged gas. It is often said, in fact, that the majority of the universe exists in the plasma phase. In addition, certain materials such as iron can become magnetic, and *ferromagnetism* is another phase of nature. Similarly, there are *liquid crystal* phases that have properties intermediate between those of liquids and solids. Nature also produces *superconducting* and *superfluid* phases that have incredible zero-resistance properties of their own. Finally, solids can come in a variety of phases with distinct crystal structures. For example, sulfur has at least two solid phases – one rhombic and the other monoclinic (see Problems 1.5–1.7 at the end of this chapter). Though not generally appreciated, water also has a number of different solid ice phases. We'll discuss the interesting and unique case of water in Chapter 3, and the other phases mentioned above will be covered in Chapters 5–7.

## 1.2 Pressure–Temperature (*P*–*T*) Phase Diagrams

Phases of matter generally occur over a range of temperatures and pressures. This is illustrated in the "typical" phase diagram shown in Figure 1.1. Here we show pressure $P$ plotted along the vertical axis, and temperature $T$ plotted along the horizontal axis. The regions corresponding to the solid, liquid, and gas phases are indicated by the corresponding labels.

Notice that the solid phase occurs in the region of the phase diagram where the pressure is high and the temperature is low, as one would expect. Similarly, the gas phase occurs in the region where the temperature is high and the pressure is low. The liquid phase occupies the wedge-shaped area of intermediate temperatures and pressures between the solid and gas phases.

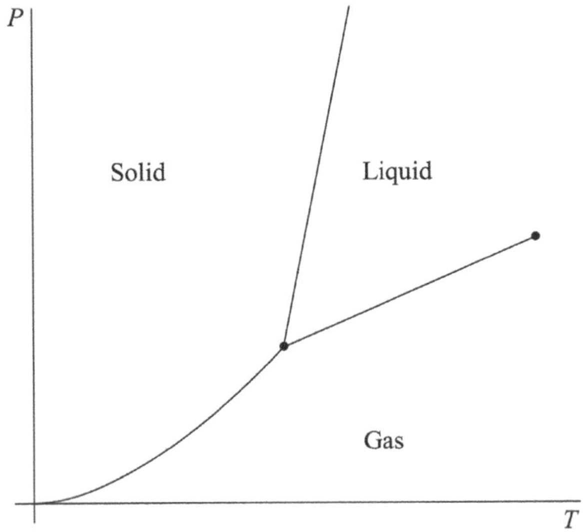


Figure 1.1 A standard pressure-versus-temperature ($P$–$T$) phase diagram for a three-phase (solid–liquid–gas) system. The solid curves indicate the location of phase transitions from one phase to another. The two black dots are discussed in Figure 1.3.

The boundaries between the phases in Figure 1.1 are the locations where the system undergoes a *phase transition*; that is, they are the locations where the system converts itself from one phase to another. The various phase transitions are indicated in Figure 1.2. For example, consider the boundary between the solid and liquid phases. If you cross this boundary in the direction of increasing temperature, the system starts off as a solid and then *melts* as it converts to the liquid phase. During the melting process, the two phases – solid and liquid – *coexist* in the system. On the other hand, if you cross in the direction of decreasing temperature, the system starts off as a liquid and then *freezes* into a solid.

Similarly, if the liquid–gas boundary is crossed in the direction of increasing temperature the liquid *vaporizes*, or *boils*, as it converts to the gas phase. Decreasing the temperature across this boundary results in the gas phase *condensing* to form a liquid. Crossing the solid–gas boundary in the direction of increasing temperature results in the solid *sublimating* into the gas phase (as in the case of dry ice, solid $CO_2$); crossing in the direction of decreasing temperature results in the gas phase *solidifying* into a solid. Each of these phase transitions is characterized by a *discontinuity* in the density of the system, with the solid phase denser than the liquid and gas phases, and the liquid phase denser than the gas phase.

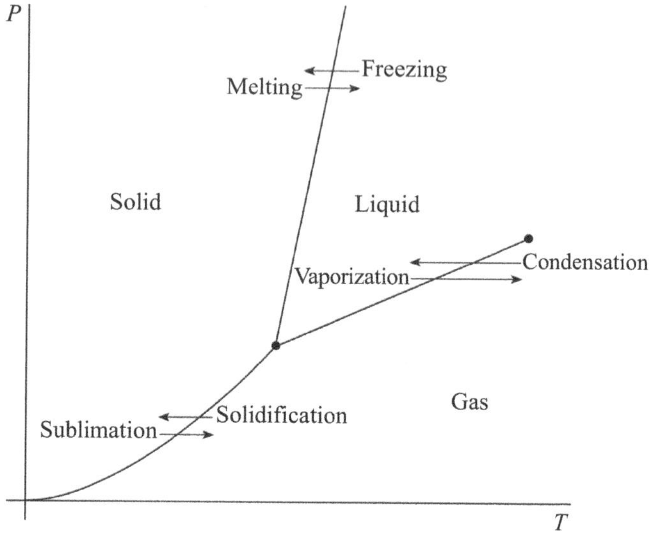


Figure 1.2 Phase transitions corresponding to each of the boundaries shown in Figure 1.1.

Now that we know the general layout of a *P–T* phase diagram, and the types of phase transitions that are possible, we turn to the two black dots that are shown in Figures 1.1 and 1.2. The dot at the intersection of the three phase boundaries is known, as one might suspect, as the *triple point.* This point is indicated in Figure 1.3. At this particular value of temperature and pressure, the three phases – solid, liquid, and gas – all coexist. In the case of water, for example, at the triple point you can have a system that is in perfect equilibrium with ice cubes floating in boiling water all bathed in steam.

The other black dot in Figure 1.3, the one at the far end of the liquid–gas phase boundary, is known as the *critical point.* At this point the distinction between liquid and gas disappears, and the system is referred to as a fluid. For example, we've mentioned that there is a discontinuity in the density of the system when crossing the liquid–gas phase boundary. The size of the discontinuity decreases as the critical point is approached. In fact, the *discontinuity vanishes* at the critical point – beyond that there is no longer a distinction between liquid and gas.

Another way to see the connection between the liquid and gas phases is shown in Figure 1.4. Here we show two points in the phase diagram labeled 1 and 2. Point 1 corresponds to a splashy liquid, and point 2 corresponds to a low-density gas. If we go from point 1 to point 2 along the path labeled A, we encounter a phase transition where the liquid boils furiously as it converts the system to the gas phase.

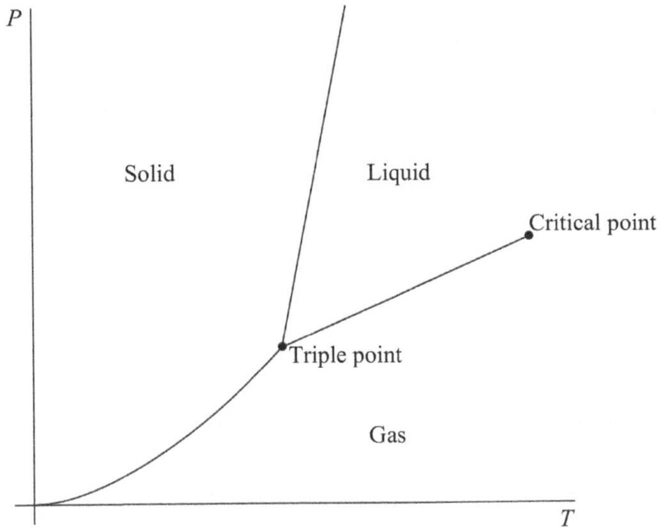


Figure 1.3 The point where the three different phase boundaries meet is known as the triple point. The solid, liquid, and gas phases coexist at this point. The high-temperature, high-pressure terminus of the liquid–gas phase boundary is known as the critical point. At this point, the distinction between the liquid and gas phases vanishes, and the system can be thought of simply as a fluid.

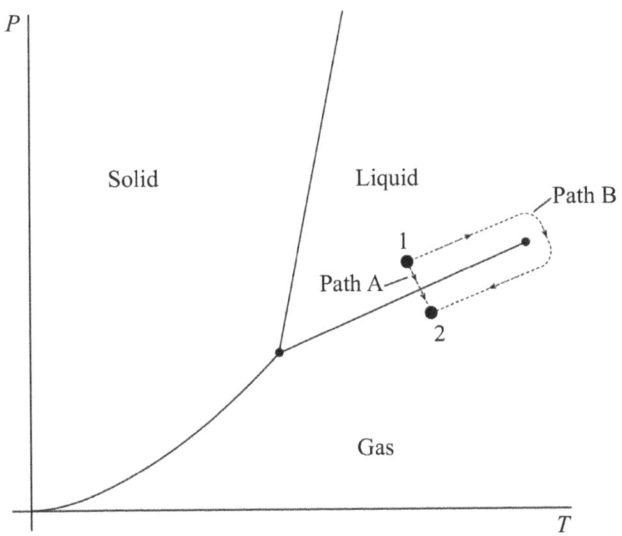


Figure 1.4 Point 1 is in the liquid phase, and point 2 is in the gas phase. If these points are connected by path A, there is a phase transition as the liquid boils away into a gas. If path B is followed, the system never undergoes a phase transition; it changes continuously from a liquid to a gas.

In contrast, suppose we go from point 1 to point 2 along the path labeled B. In this case, we first increase the pressure and temperature, then continue to increase the temperature as we decrease the pressure, and finally decrease both the pressure and temperature until we arrive at point 2. The beginning and ending points are precisely the same as before, but this time the system never undergoes a phase transition. The density of the system simply changes in a continuous way along the path until the liquid is now a gas. In this sense, we say that there is no fundamental distinction between a liquid and a gas – they are one and the same phase, the fluid phase, and whether we see a phase transition between them or not depends on our path in parameter space. There is no corresponding critical point between the solid phase and the other phases.

To enhance our understanding of *P–T* phase diagrams, let's consider various paths through the diagrams. For example, in Figure 1.5 we see three paths of constant pressure, as indicated by the dashed numbered lines. Line 1 corresponds to a high pressure. If we start on this line at low temperature the system is in the solid phase. Raising the temperature brings us to a phase transition where the solid melts to a liquid.

Similarly, consider line 2. Starting once again at a low temperature, the system begins in the solid phase. As we raise the temperature, the system encounters the solid–liquid phase boundary, and the solid melts to a liquid.

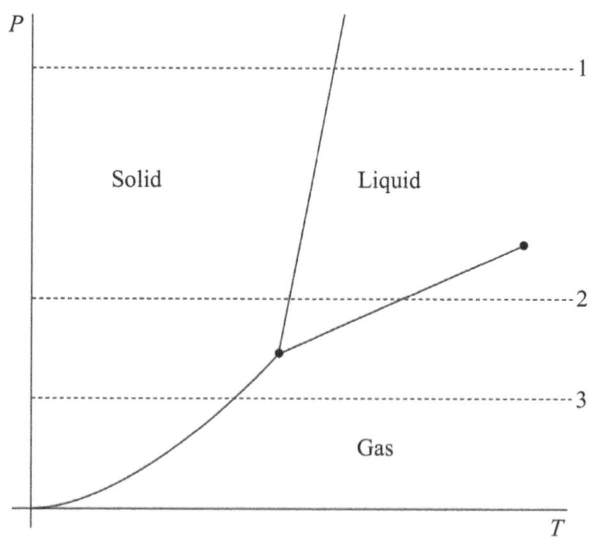


Figure 1.5 A *P–T* phase diagram with three dashed lines of constant pressure.

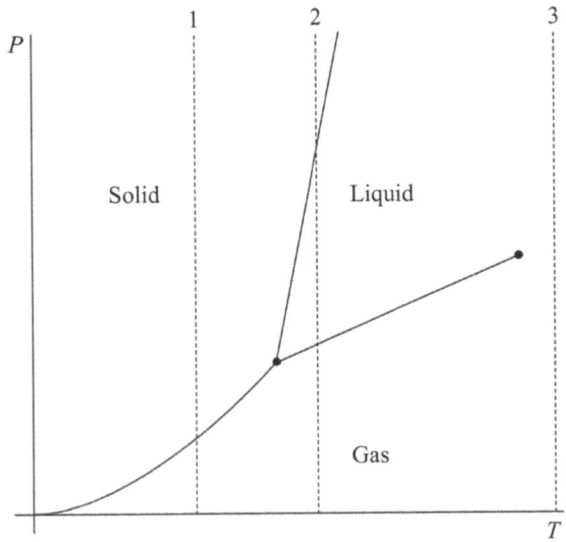


Figure 1.6 A *P–T* phase diagram with three dashed lines of constant temperature.

At a slightly higher temperature the liquid–gas phase boundary is encountered, and the liquid boils into the gas phase. Finally, on line 3 the system undergoes a sublimation phase transition from solid to gas.

In Figure 1.6 we see the same *P–T* phase diagram, but this time with three dashed lines of constant temperature. On line 1 the system is a gas at low pressure, but as the pressure is increased the gas undergoes a phase transition to the solid phase – the opposite of sublimation, sometimes referred to as *deposition*. Line 2, at an intermediate temperature, again starts in the gas phase at low pressure. As the pressure is increased, the gas condenses as the system undergoes a phase transition to the liquid phase. At a still higher pressure, the liquid solidifies. At the highest temperature, corresponding to line 3, the system is in the gas (or fluid) phase for all pressures shown in the phase diagram. Eventually, at still higher pressures, the gas will solidify.

Now that we've explored the *P–T* phase diagrams, we turn to some of the underlying physics behind the phase transitions.

## 1.3 Gibbs Free Energy

One of the basic principles of statistical mechanics is that a system in equilibrium tends to minimize its free energy. In the present case, where the phase diagram is plotted on a $P$-versus-$T$ plane, the free energy of interest is the *Gibbs free energy*, $G$. This is because the independent variables for $G$ are pressure, $P$, and temperature, $T$. We show this aspect of $G$ in this section, as well as how to use $G$ to determine the slope of phase boundaries.

## Energy Conservation

We start with the basic statement of energy conservation in a thermodynamic system. Letting $dE$ stand for the change in energy of a system (another common choice is $dU$), and with $dQ$ representing the exchange of heat energy and $dW$ the mechanical work done on the system, we have

$$dE = dQ + dW.$$

Let's write this out in terms of thermodynamic quantities.

We'll start with the heat exchange, which can be written as

$$dQ = TdS.$$

This relates heat and absolute temperature to the change in entropy, $dS$; for example, adding heat to a system increases its entropy. Similarly, for the work term, we have

$$dW = -PdV.$$

The negative sign accounts for the fact that a pressure acting to compress a system (change in volume, $dV < 0$) increases the energy of the system ($dW > 0$). Combining these relationships, we find that

$$dE = TdS - PdV. \tag{1.1}$$

This is energy conservation – the first law of thermodynamics – stated in terms of thermal quantities.

Notice that $E$ depends on the independent variables $S$ and $V$:

$$E = E(S, V).$$

That is, the differential $dE$ is related to the differential changes $dS$ and $dV$. Specifically, temperature and pressure are related to $E$ as follows:

$$T = \left.\frac{\partial E}{\partial S}\right|_V,$$

$$P = -\left.\frac{\partial E}{\partial V}\right|_S.$$

These are not the most useful variables, however, because $S$ is not readily measurable. As a result, we now turn to transforming the energy to other forms that depend on more convenient variables.

### Legendre Transformations

We start by transforming $E(S,V)$ to a new function that depends on $T$ and $V$ rather than $S$ and $V$. This type of change in variables is referred to as a *Legendre transformation*. Basically, we'd like to change the $TdS$ term in Eq. (1.1) to a term involving $SdT$, so that $T$ is now the independent variable.

We do this by introducing the Helmholtz free energy, $F$, defined as

$$F = E - TS.$$

To see the utility of this definition, consider the differential of $F$. We find that

$$dF = dE - d(TS). \tag{1.2}$$

Writing out the second term using the chain rule yields

$$d(TS) = TdS + SdT.$$

Substituting this result in Eq. (1.2) gives

$$\begin{aligned} dF &= dE - (TdS + SdT) \\ &= (TdS - PdV) - (TdS + SdT) \\ &= -SdT - PdV. \end{aligned} \tag{1.3}$$

By canceling the $TdS$ terms, we see that the independent variables for $F$ are now $T$ and $V$, which are much more convenient for measurement. Specifically,

$$F = F(T,V).$$

It follows from Eq. (1.3) that

$$S = -\left.\frac{\partial F}{\partial T}\right|_V,$$

$$P = -\left.\frac{\partial F}{\partial V}\right|_T.$$

Thus, the process of subtracting $TS$ has changed the independent variable $S$ to the independent variable $T$.

To change the $V$ dependence to a $P$ dependence, we perform another Legendre transformation, this time using a $PV$ term. Thus, we define the Gibbs free energy, $G$, as

$$G = F + PV.$$

Taking the differential, we find that

$$dG = dF + d(PV) = (-SdT - PdV) + (PdV + VdP).$$

Canceling the $PdV$ terms yields

$$dG = -SdT + VdP. \tag{1.4}$$

Thus, $G = G(T, P)$is a function of $T$ and $P$, as desired, and

$$S = -\left.\frac{\partial G}{\partial T}\right|_P,$$

$$V = \left.\frac{\partial G}{\partial P}\right|_T.$$

These are particularly convenient variables, since $P$ and $T$ are readily measured with common devices.

## The Gibbs Free Energy and Phase Transitions

We return now to the principle stated at the beginning of this section – namely, that a system in equilibrium minimizes its free energy. Specifically, we'll apply this principle to the Gibbs free energy $G$ in terms of the variables $T$ and $P$.

To begin, consider Figure 1.7, which shows a schematic plot of the Gibbs free energy versus temperature for a constant pressure. In this system there are two distinct phases, labeled phase 1 and phase 2. This plot corresponds to line 1 in Figure 1.5. Thus, at low temperature the system is in the solid phase, represented by phase 1 in Figure 1.7; at high temperatures the system is a liquid, represented by phase 2.

Now, these two phases are quite distinct in their physical characteristics, and hence we expect that $G$ will depend quite differently on the temperature in them, as shown in Figure 1.7. At some temperature, however, $G$ for phase 1 intersects with $G$ for phase 2. It follows that the minimum value of $G$ switches from one phase to the other at this temperature – which means that the system undergoes a phase transition. The dot-dashed lines indicate unstable, or more accurately *metastable*, extensions of the two phases. By metastable, we mean that the dot-dashed lines are the location of *local* minima of the free energy, as opposed to the *global* minima represented by the solid lines.

The value of $G$ is continuous at the phase transition, but its first derivative is discontinuous. For this reason, older literature often refers to a phase transition like this as a "first-order" phase transition. It is more common in modern usage to simply refer to this transition as "discontinuous" because the discontinuity

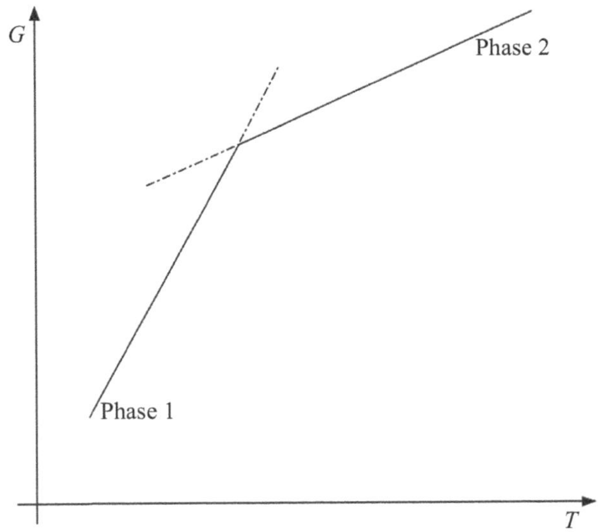


Figure 1.7 A schematic plot showing the Gibbs free energy, $G$, versus temperature $T$ for constant pressure. Note that there are two distinct phases in this system, labeled phase 1 and phase 2. In equilibrium, the system is in the phase that minimizes the free energy, as indicated by the continuous dark lines. It follows that a phase transition from phase 1 to phase 2 occurs at the intersection point. The dashed lines indicate metastable extensions of the two phases.

in the first derivative implies a discontinuity in certain physical quantities, such as the density and entropy, for example.

Next, we consider the intermediate pressure case in Figure 1.5, as indicated by line 2. At a low temperature the system is a solid, as represented by phase 1 in Figure 1.8. As the temperature is raised, the solid melts to the liquid phase (phase 2); as the temperature is raised further, the liquid boils to the gas phase (phase 3). Each of these phase transitions is discontinuous, as indicated by the kinks in Figure 1.8. The dot-dashed lines indicate metastable extensions of the three phases.

Finally, Figure 1.7 also corresponds to line 3 in Figure 1.5, though in this case the phases are different. Specifically, at a low temperature the system is in the solid phase (phase 1), and at high temperature the system is in the gaseous phase (phase 2). Other than that, the interpretation of Figure 1.7 is the same.

## The Clausius–Clapeyron Relation

We now show how the Gibbs free energy can be used to determine the slope of a phase boundary in a $P$–$T$ plot. This result is very useful in studying phase diagrams, and we will return to it in more detail in the next two chapters.

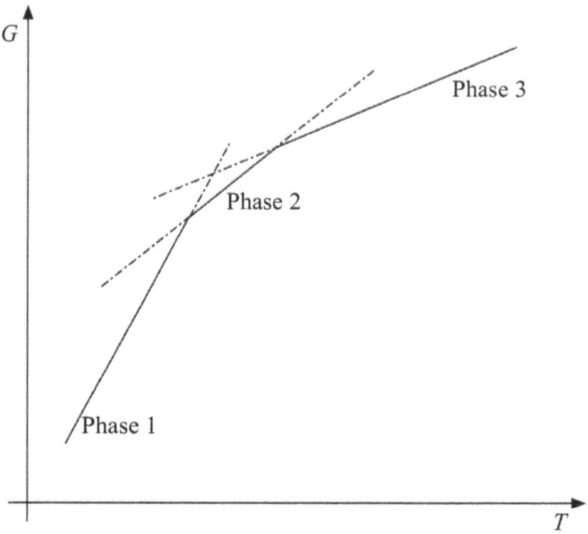


Figure 1.8 A schematic plot showing the Gibbs free energy $G$ versus temperature $T$ for constant pressure. Note that there are three distinct phases in this system, labeled phase 1, phase 2, and phase 3. In equilibrium, the system is in the phase that minimizes the free energy, as indicated by the continuous dark lines. The dot-dashed lines indicate metastable extensions of the three phases.

To begin, consider the enlarged portion of a phase boundary shown in Figure 1.9. The two phases, phase 1 and phase 2, are in equilibrium along this boundary. For example, at point i the Gibbs free energy in phase 1, $G_1$, is equal to the Gibbs free energy in phase 2, $G_2$; that is,

$$G_1 = G_2.$$

Now, as we move along the phase boundary from point i to point f, the change in the Gibbs free energy of phase 1 is

$$\Delta G_1 = -S_1 \Delta T + V_1 \Delta P.$$

Similarly, the change in the Gibbs free energy in phase 2 is

$$\Delta G_2 = -S_2 \Delta T + V_2 \Delta P.$$

Now, if point f is to still be on the phase boundary, it is necessary that the change in $G$ for phase 1 must be equal to the change in $G$ for phase 2. That is,

$$\Delta G_1 = \Delta G_2.$$

This is the basic condition defining the phase boundary.

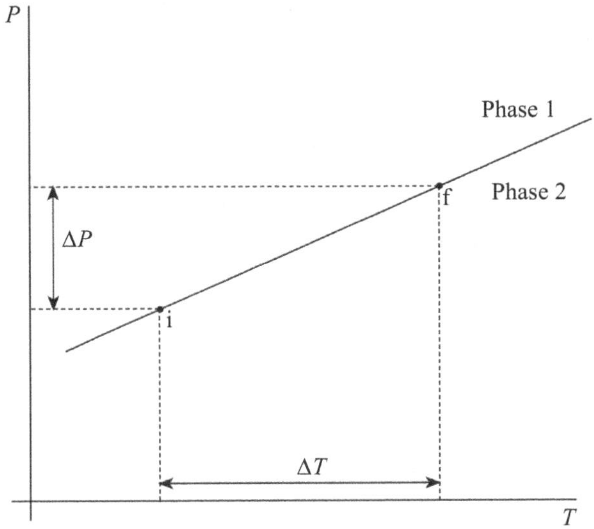


Figure 1.9 Enlarged section of a phase boundary between phase 1 and phase 2. The change in pressure $\Delta P$ and change in temperature $\Delta T$ on the phase boundary are related by equating the change in Gibbs free energy from point i to point f.

Using the results just given, the condition $\Delta G_1 = \Delta G_2$ can be written as

$$-S_1\Delta T + V_1\Delta P = -S_2\Delta T + V_2\Delta P.$$

Rearranging yields

$$(V_1 - V_2)\Delta P = (S_1 - S_2)\Delta T.$$

Solving this for the rate of change of pressure with temperature, and going to the infinitesimal limit, we find that

$$\frac{dP}{dT} = \frac{(S_1 - S_2)}{(V_1 - V_2)}. \tag{1.5}$$

This result is known as the *Clausius–Clapeyron relation.*

Notice that the Clausius–Clapeyron relation gives the slope of the phase boundary in a *P–T* diagram in terms of the change in entropy, $\Delta S = (S_1 - S_2)$, and the change in volume, $\Delta V = (V_1 - V_2)$, between two phases at a given point. As already mentioned, the entropy is not readily measurable, and so this relation is not particularly useful as it stands. A more useful form, expressed in terms of the heat associated with a phase transition, will be presented in Chapter 2.

Even so, we can glean some useful and interesting qualitative information from the Clausius–Clapeyron relation in its current form. For example, consider the phase boundary between a solid phase, phase 1, and a gas phase, phase 2. Clearly, the entropy of the gas phase – with molecules occupying random positions – is greater than the entropy of the solid phase –with molecules in precise locations. It follows that $(S_1 - S_2) < 0$, and so the numerator of Eq. (1.5) is negative. Similarly, the volume of a given mass of the solid is less than the volume of an equal mass of the gas. Therefore, $(V_1 - V_2) < 0$ and the denominator of Eq. (1.5) is also negative. As a result, it follows that $\frac{dP}{dT} > 0$ for the solid–gas phase boundary, as we see in the phase diagrams in Figures 1.1–1.6.

In contrast, consider the phase diagram for water. In this case, let phase 1 be ice and phase 2 be water. The entropy of liquid water is greater than that of solid ice, and hence $(S_1 - S_2) < 0$. However, the volume of a given mass of ice is great than that of water – which is why ice floats on water, of course. Thus $(V_1 - V_2) > 0$, and it follows that $\frac{dP}{dT} < 0$. In Figures 1.1–1.6 we show a positive slope for the solid–liquid phase boundary, but this slope will be negative for the exceptional case of water, as we shall see in greater detail in Chapter 3.

## 1.4 Pressure–Volume (*P–V*) Phase Diagrams

Additional insight into phase transitions can be gained by plotting the standard phase diagram in terms of different variables. So far, we've considered the phase diagram on a pressure-versus-temperature (*P–T*) plot, as in Figure 1.10. We now explore the corresponding phase diagram on a pressure-versus-volume (*P–V*) plot, as shown in Figure 1.11. Let's explore the connection between these phase diagrams.

To begin, note that the phase boundaries in Figure 1.10 correspond to locations where two different phases are in *coexistence* (or three phases at the triple point, of course). These phases have different densities – think of water and steam, for example – which means that the volume *V* occupied by a given amount of material is different as well. The amount of this difference in volume is indicated qualitatively by the length of the double-headed arrows in Figure 1.10.

For example, the lowest double-headed arrow indicates the rather large difference in volume between the solid phase and a coexisting gaseous phase. The difference in density between a coexisting solid and liquid phase is rather small in comparison, as indicated by the uppermost double-headed arrow. Along the coexistence line separating liquid and gas, we see that the difference in volume is fairly large near the bottom of the line and decreases as the critical

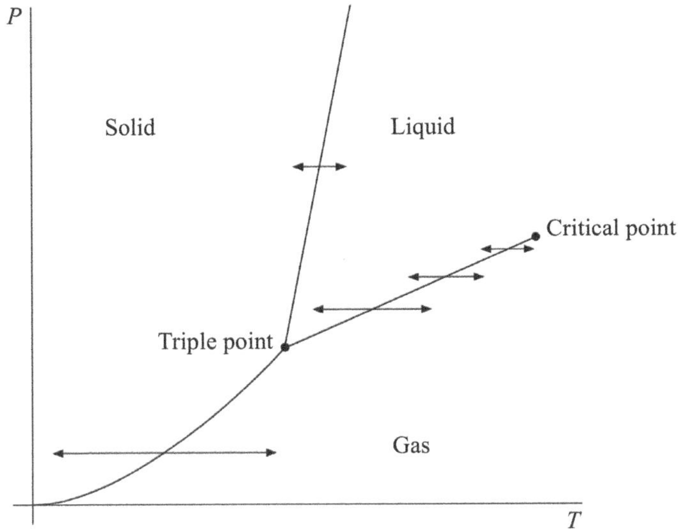


Figure 1.10 Standard *P–T* phase diagram. The double-headed arrows are qualitative indications of the difference in volume of the coexisting phases along each phase boundary.

point is approached. Of course, at the critical point itself, the difference in volume, or density, vanishes.

With these observations in mind, we can convert the temperature axis in Figure 1.10 to a volume axis by splitting each coexistence line into two coexisting volumes. The result is shown in Figure 1.11. First, we see the large difference in volume between the coexisting solid and gas phases, as indicated by the horizontal dashed lines across the (solid + gas) portion of the phase diagram. Similarly, the rather small difference in volume between the coexisting solid and liquid phases is indicated by the dashed lines across the (solid + liquid) portion of the phase diagram. The decreasing difference in volume between the coexisting liquid and gas phases, and the eventual vanishing of this difference at the critical point, is indicated by the dashed lines across the (liquid + gas) portion of the phase diagram. Finally, the triple point in the *P–T* phase diagram has been separated into three different volumes in the *P–V* phase diagram to form a *triple line*.

One of the advantages of the phase diagram in Figure 1.11 is that we can easily read off from it the behavior of the system at constant volume, *V*, or at constant pressure, *P*. For example, suppose we start at high pressure with the volume corresponding to the critical point. The system is a fluid at this pressure, with no distinction between liquid and gas. As we lower the pressure,

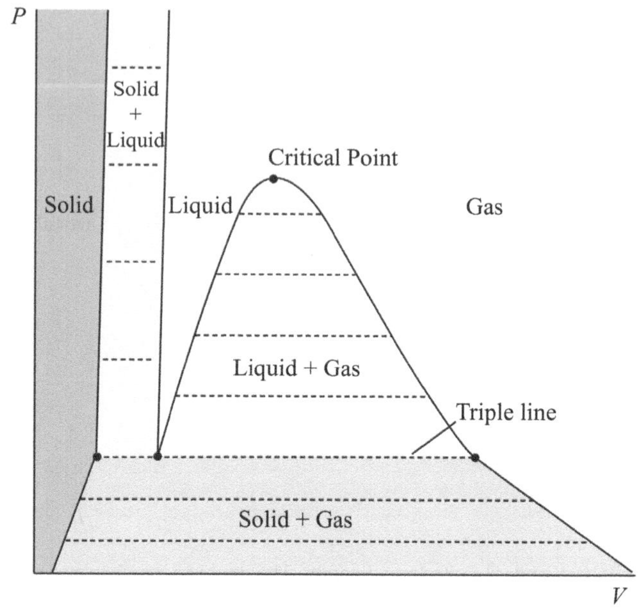


Figure 1.11 Standard phase diagram converted to pressure-versus-volume, $P$–$V$. The horizontal dashed lines indicate the two volumes, or densities, that coexist at a given pressure. The exception is the triple line, where three different volumes coexist.

by reducing the temperature for example, the system separates into coexisting liquid and gas phases, with the difference in volume of the two phases increasing with decreasing pressure. At the triple point pressure, the solid phase – the third phase of the system – enters the picture. Below this pressure we once again have just two phases, this time with the solid and gas in coexistence.

Similarly, suppose we start at a constant pressure between that of the triple line and the critical point and vary the volume of the system. At large volume the system is a fluid. As we compress the system the gas phase condenses to a liquid, and these phases coexist. Decreasing the volume more results in a pure liquid phase. Finally, at still smaller volumes, the solid phase appears, first in coexistence with the liquid, and then as the only phase in the system.

## Isotherms in *P–V* Phase Diagrams

Even more information can be obtained by considering the behavior of the system along curves of constant temperature. These are referred to as *isotherms*, from the root words *iso* (equal) + *therm* (temperature). We show

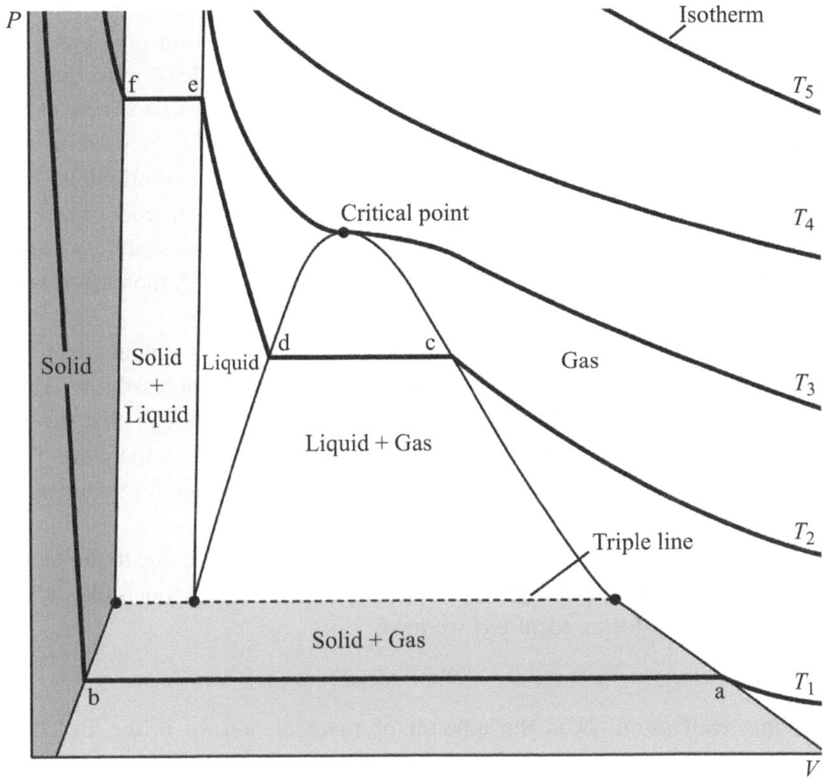


Figure 1.12 Isotherms on a $P$–$V$ phase diagram for equally spaced increasing temperatures from $T_1$ to $T_5$. Points a and b show the coexistence of solid and gas phases; points c and d show the coexistence of liquid and gas phases; points e and f show the coexistence of solid and liquid phases.

five different isotherms in Figure 1.12, labeled $T_1$ to $T_5$, corresponding to increasing temperature.

First, consider the lowest-temperature isotherm, $T_1$. Imagine we have a finite amount of material in a container, and we compress it to a smaller volume, all the time keeping the temperature constant. At high volume the system is a gas, and decreasing the volume increases the pressure, as we see on curve $T_1$ as it approaches point a. When the system reaches point a, the solid phase begins to appear. *Compressing the system further results in no change in pressure, but an increasing amount of solid.* This continues until the entire system has been converted to solid, at point b. Further compression results in a sharp rise in pressure, due to the low compressibility of the solid.

Next, consider the isotherm $T_2$. Once again, at high volume the system is a gas, and compressing the system results in a steady rise in pressure – until point c is reached. At this point the gas begins to condense into the liquid phase. *Further decreasing the volume results in no change in pressure until all of the gas has condensed to liquid*, at point d. Additional compression results in a sharp rise in pressure – since the liquid is almost incompressible – until point e is reached, at which point the solid phase appears in coexistence with the liquid. *Reducing the volume further converts all the liquid to solid at constant pressure*, at point f. Beyond this point further compression results in a rapid rise in pressure.

At temperature $T_3$ the isotherm just touches the critical point, at which point it is momentarily horizontal – which implies infinite compressibility. Hence, the system remains a fluid until the pressure is great enough that the solid phase begins to appear, out of the range of this plot. At temperature $T_4$ the isotherm still shows a bit of a flattening, but other than that it remains a fluid until the solid begins to form at high pressure.

Finally, at $T_5$ the isotherm has a simple monotonic shape due to the fact that at high temperature and high volume the system acts very much like an ideal gas. Recall that for an ideal gas we have

$$PV = Nk_{\mathrm{B}}T.$$

In this expression, $N$ is the amount of material and $k_B$ is the Boltzmann constant. Solving for $P$ as a function of $V$ on an isotherm, we have

$$P = \frac{Nk_{\mathrm{B}}T}{V} = \frac{\text{constant}}{V}.$$

This is the expression for a hyperbola. Thus, as expected, the high-temperature isotherms, like $T_5$, have the basic shape of a hyperbola.

We will explore this phase diagram, and the *P–T* phase diagram, in more detail in Chapter 2. We will also discuss the amount of heat that is associated with each of the various phase transitions.

## 1.5 Problems

1.1 **Densities.** Figure 1.13 shows a typical *P–T* phase diagram.

**(a)** Is the density of the system at point 1 greater than, less than, or equal to the density at point 2? Explain why.

**(b)** Is the density at point 3 greater than, less than, or equal to the density at point 2? Explain why.

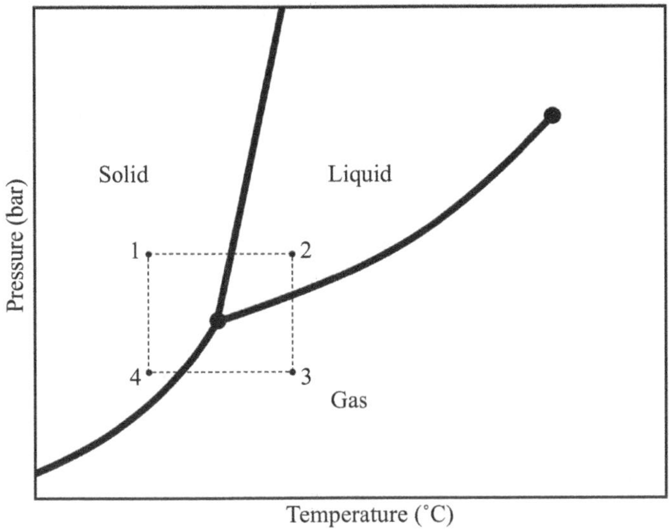


Figure 1.13 Typical *P–T* phase diagram with four designated points. This figure applies to Problems 1.1 and 1.2.

1.2 **Phase transitions.** Referring to Figure 1.13, describe **(a)** the phase transitions in going from point 1 to point 2 to point 3 and **(b)** the phase transitions in going from point 1 to point 4 to point 3.

1.3 **Densities.** Referring to the phase diagram in Figure 1.14, is the density at point 1 greater than, less than, or equal to the density at point 3? Explain why.

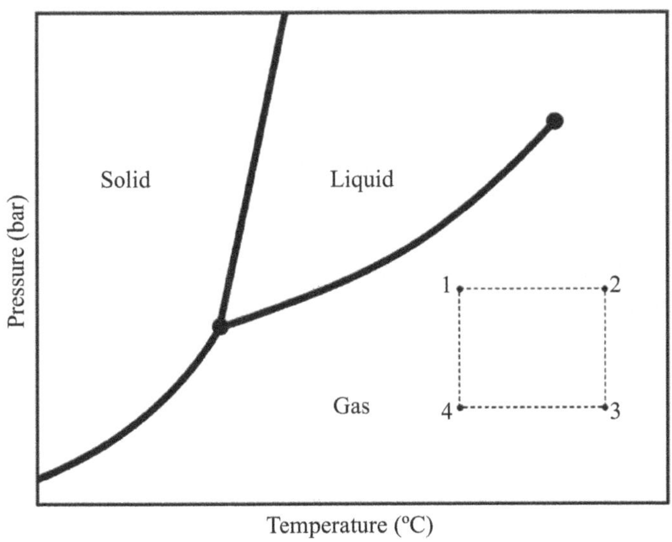


Figure 1.14 Typical *P–T* phase diagram with four designated points. This figure applies to Problem 1.3.

1.4 **Interpreting phase diagrams.** Refer to the phase diagram in Figure 1.15.
**(a)** Is the density at point 1 greater than, less than, or equal to the density at point 3? Explain why.
**(b)** Describe the phase transitions in going from point 1 to point 2 to point 3.
**(c)** Describe the phase transitions in going from point 1 to point 4 to point 3.

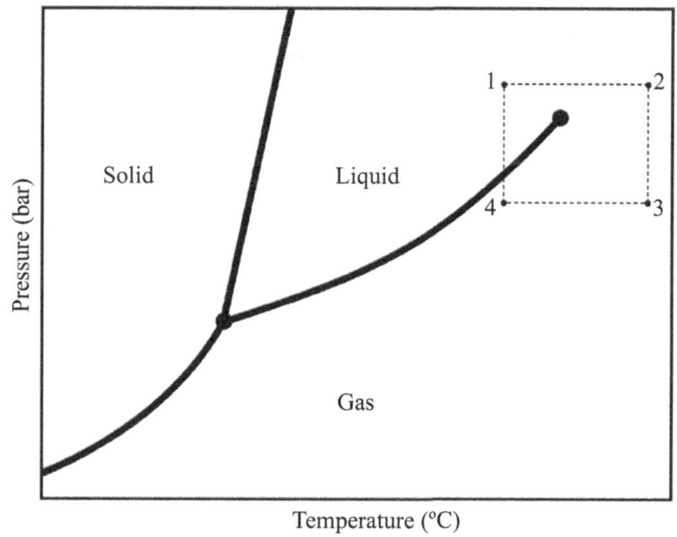


Figure 1.15 Typical *P–T* phase diagram with four designated points. This figure applies to Problem 1.4.

1.5 **Interpreting phase diagrams.** The phase diagram for sulfur is given in Figure 1.16. In this phase diagram, solid 1 is a rhombic phase and solid 2 is a monoclinic phase. How many triple points are there in this system?

1.6 **Interpreting phase diagrams.** Referring to Figure 1.16, describe the sequence of phase transitions as temperature is increased at a pressure of 1 atmosphere.

1.7 **Interpreting phase diagrams.** Referring to Figure 1.16, describe the sequence of phase transitions as pressure is increased at a temperature of 114 °C.

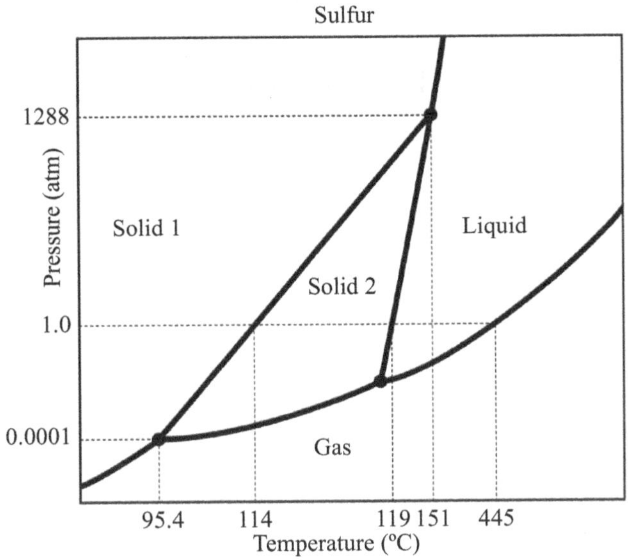


Figure 1.16 Phase diagram for sulfur. Solid 1 is a rhombic phase, and solid 2 is a monoclinic phase.

# 2

# Heat and Phases

A key aspect of phases and phase transitions is heat. For example, if we add heat to a system its entropy increases. The added heat may or may not increase the temperature, however. In addition, large quantities of heat can be absorbed or released when a system undergoes a transition from one phase to another. All of these characteristics of heat and phases, and more, are explored in detail in this chapter.

## 2.1 Heat, Entropy, and Temperature

When heat is added to a system, one thing we know for sure is that its entropy increases. In a reversible process, the differential relationship between heat $Q$ and entropy $S$ is

$$dQ = TdS. \tag{2.1}$$

In this expression, $T$ is the absolute temperature, measured in kelvins, a quantity that is always positive. It follows that adding heat to a system, which means $dQ > 0$, results in an increase in entropy, $dS > 0$. The increase in entropy is even greater if the process is irreversible.

For example, suppose you add heat to the air contained in a balloon. This causes the air molecules to move more rapidly, resulting in an expansion of the balloon. Clearly, the entropy of the system has increased. If the balloon pops, and releases the air it contained – an irreversible process – the increase in entropy is even greater. Similarly, adding heat to a crystal causes the molecules to move more rapidly, and with greater amplitude, about their equilibrium positions. This also results in an increase in entropy. On the other hand, withdrawing heat from a system reduces its entropy.

One might think that similar remarks apply to the temperature of a system. There's a fundamental difference, however. To see this, suppose we put a pan of water on the stove and add heat to it. Clearly, the temperature of the water increases. This increase in temperature continues until the boiling temperature of 100 °C is reached. At this point, the water begins to bubble and boil – it undergoes a phase transition – and steam rises from the pan. Adding more heat to the system *does not* result in an increase in temperature, however. Instead, it simply causes more steam to boil off from the water. This continues until all of the water is converted to steam. Only then can the temperature of the system rise to a value greater than 100 °C.

In Section 2.2 we consider the case of heating a substance that is in a given phase. Both the entropy and temperature increase with added heat in such a process. In Section 2.3 we consider the heat added to a system to cause a phase transition; that is, to convert the substance from one phase to another. As mentioned, this process occurs at constant temperature.

## 2.2 Heat Capacity and Specific Heat

If we add the quantity of heat $\Delta Q > 0$ to a substance, and its temperature rises by the amount $\Delta T > 0$, we define the *heat capacity, C,* of the substance as

$$C = \frac{\Delta Q}{\Delta T}. \tag{2.2}$$

The heat capacity is always positive, and its dimensions are joules per kelvin (J/K). Equivalently, since degrees have the same size for the Kelvin and Celsius scales, we could also use joules per degree Celsius (J/°C).

As an example, suppose we have a system that consists of 1 kg of water. To raise the temperature of this system by 1 K, we must add 4186 J of heat. If, instead, we want to increase the temperature of 1 kg of gold by 1 K we need only add 128 J of heat. Thus, a given mass of each *type* of substance has its own *characteristic* heat capacity.

The examples mentioned above refer to 1 kg of substance. Clearly, however, the amount of heat needed to yield a given temperature increase is proportional to the amount of material in the system – that is, to the system's mass $m$. For example, to raise the temperature of 2 kg of water by 1 K requires $2 \times (4186\,\text{J})$. Thus, it is useful to talk about the *heat capacity per mass* because this quantity is a characteristic of the substance and doesn't depend on the amount of substance in the system. We refer to this quantity as the **specific heat capacity**, $c$, and define it as

Table 2.1 *Specific heats for a variety of substances at atmospheric temperature and pressure*

| Substance | Specific heat $c$ [J/(kg·K)] |
|---|---|
| Water | 4186 |
| Ice | 2090 |
| Steam | 2010 |
| Air | 1004 |
| Aluminum | 900 |
| Glass | 837 |
| Iron (steel) | 448 |
| Copper | 387 |
| Silver | 234 |
| Gold | 129 |
| Lead | 128 |

*Note:* Data taken from Walker, James S. *Physics*, 5th ed. (Addison-Wesley, 2016).

$$c = \frac{\Delta Q}{m\Delta T}. \tag{2.3}$$

The dimensions of the specific heat capacity, or *specific heat* for short, are $\mathrm{J/(kg \cdot K)}$.

The specific heats for a variety of substances are presented in Table 2.1. We note immediately that the value for water is much greater than those for the other substances. This is just one of the many striking characteristics that set water apart from other common substances. We explore water in detail in Chapter 3.

The specific heats given in Table 2.1 are for atmospheric temperature and pressure. These values will change with temperature and pressure, but we can consider the specific heat to be constant for the temperature and pressure ranges of many ordinary circumstances.

Recall that a large specific heat, like that for water, means that a lot of heat is required to change the temperature. On the other hand, a small specific heat, like that for gold, means that a given amount of heat results in a larger change in temperature.

To help visualize these differences, we rearrange Eq. (2.3) to give

$$\Delta T = \frac{\Delta Q}{mc}.$$

Going to the differential limit, and integrating both sides of the equation, yields

$$\int dT = \int \frac{1}{mc} dQ.$$

Now, let's integrate from an initial temperature $T_0$, with no heat added, to a final temperature $T$, with the heat $Q$ added. The corresponding integrals are

$$\int_{T_0}^{T} dT = \int_{0}^{Q} \frac{1}{mc} dQ. \tag{2.4}$$

The temperature integral is

$$\int_{T_0}^{T} dT = T - T_0.$$

Assuming the mass $m$ and specific heat $c$ are constant over the desired temperature range, the heat integral is

$$\frac{1}{mc} \int_{0}^{Q} dQ = \frac{1}{mc}(Q - 0).$$

Substituting these results into Eq. (2.4) and rearranging yields

$$T = T_0 + \left(\frac{1}{mc}\right) Q.$$

This result is plotted in Figure 2.1 for $T_0 = 10.0\,°\mathrm{C}$, $m = 1.00\,\mathrm{kg}$, and various values for $c$, depending on the substance.

Notice that the temperature of water barely rises over this range of heat. In contrast, the temperature of gold increases rapidly. As we can see, the rate of increase of temperature with heat is *inversely proportional* to the specific heat.

## Integral Relations for the Heat Capacity

There are two integral relationships for the heat capacity of a substance that are quite useful, both theoretically and experimentally. First, we start with the differential version of Eq. (2.2):

$$C = \frac{dQ}{dT}.$$

Rearranging and integrating yields

$$\int C\, dT = \int dQ = \Delta Q. \tag{2.5}$$

Thus, the area under a heat capacity curve between two temperatures, given by the integral on the left-hand side of Eq. (2.5), is equal to $\Delta Q$, the heat required to produce that change in temperature.

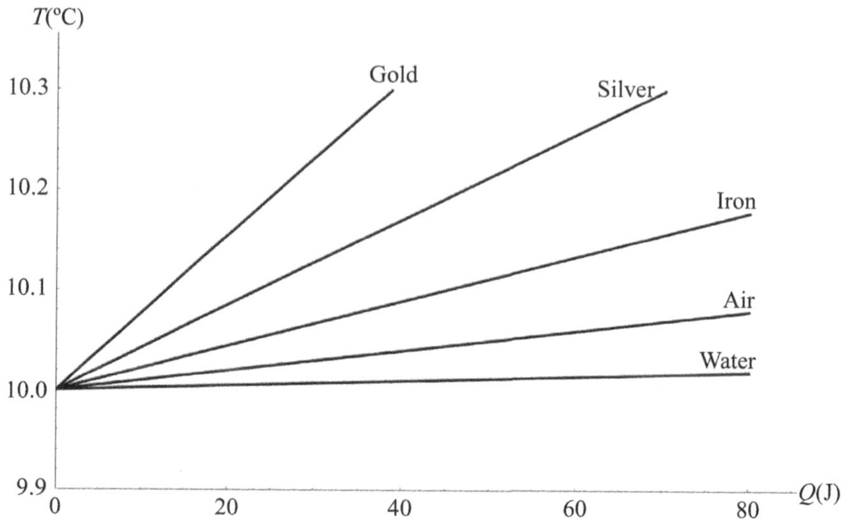


Figure 2.1 Temperature $T$ of various substances, all with an initial temperature of 10.0 °C, as a function of the heat added $Q$.

Another useful relationship is obtained by using Eq. (2.1) to express the heat in terms of entropy. This gives

$$C\,dT = dQ = T\,dS.$$

Dividing both sides by the absolute temperature $T$ and integrating yields

$$\int \frac{C}{T}\,dT = \int dS = \Delta S.$$

Thus, integrating $C/T$ over a range of temperatures gives the corresponding change in entropy. For example, if the heat capacity is constant over a given range of temperatures, the integral of $C/T$ gives

$$C \ln T + \text{constant} = \Delta S.$$

This shows the origin of the characteristic logarithmic dependence of entropy on temperature, as seen in so many systems.

## 2.3 Latent Heat

We've seen a number of examples where temperature increases with added heat. Now we turn to cases where heat added to a system results in *no change* in temperature at all – only an increase in entropy.

To be specific, consider a system of mass $m$ undergoing a phase transition. An amount of heat $\Delta Q$ must be added to complete the conversion of the system from one phase to the other. In such a case, we define the *latent heat, L,* to be the following:

$$L = \Delta Q.$$

The dimensions of the latent heat are joules, J. In analogy with the case of heat capacity and specific heat capacity, we define the *specific latent heat, l,* to be the latent heat per mass. Thus,

$$l = \frac{L}{m} = \frac{\Delta Q}{m}.$$

The dimensions of the specific latent heat are joules per kilogram, J/kg.

When the phase transition is from solid to liquid, as for ice melting to liquid water, for example, the corresponding specific latent heat is referred to as the *specific latent heat of fusion*, and is designated as follows:

$$l_\mathrm{f} = \text{specific latent heat of fusion (freezing/melting)}.$$

The name *fusion* may seem a bit odd, but it refers to the fact that as the temperature is lowered the liquid "fuses" to form a solid. Conversely, when a system goes from liquid to gas, the corresponding specific latent heat is referred to as the *specific latent heat of vaporization*:

$$l_\mathrm{v} = \text{specific latent heat of vaporization (vaporization/condensation)}.$$

Table 2.2 gives specific latent heats for a variety of substances. We don't list specific latent heats of sublimation because those transitions are relatively uncommon.

Let's take a look at a particular example to get a feeling for the role of the specific latent heat in a system that is being heated. Suppose we have 1.00 kg of ice cubes in a bowl, and that, initially, the ice cubes are at a temperature of −20.0 °C. We now add heat until the system is converted to liquid water at 20.0 °C. This process is illustrated in Figure 2.2.

The first step in the process, indicated by the line connecting points A and B, corresponds to warming the ice cubes from −20.0 °C to 0.00 °C. The heat required for this step, $Q_\mathrm{AB}$, is given in terms of the specific heat for ice, as found in Table 2.1. Thus, we have

$$Q_\mathrm{AB} = mc_\mathrm{ice}\Delta T = (1.0\ \mathrm{kg})[2090\ \mathrm{J/(kg \cdot {}^\circ C)}](20.0\ {}^\circ\mathrm{C}) = 41{,}800\ \mathrm{J}.$$

With this amount of heat, the ice cubes are brought to the melting temperature of 0 °C, and ice and water are now in equilibrium.

Table 2.2 *Specific latent heats of fusion and vaporization for various substances*

| Substance | Specific latent heat of fusion, $L_f$ (J/kg) | Specific latent heat of vaporization, $L_v$ (J/kg) |
|---|---|---|
| Ammonia | 332,000 | 1,370,000 |
| Benzene | 126,000 | 394,000 |
| Copper | 207,000 | 4,730,000 |
| Ethyl alcohol | 108,000 | 855,000 |
| Gold | 62,800 | 1,720,000 |
| Lead | 23,200 | 859,000 |
| Nitrogen | 25,700 | 200,000 |
| Oxygen | 13,900 | 213,000 |
| Silver | 88,200 | 2,330,000 |
| Water | 335,000 | 2,260,000 |

*Note:* Data taken from Walker, James S. *Physics*, 5th ed. (Addison-Wesley, 2016).

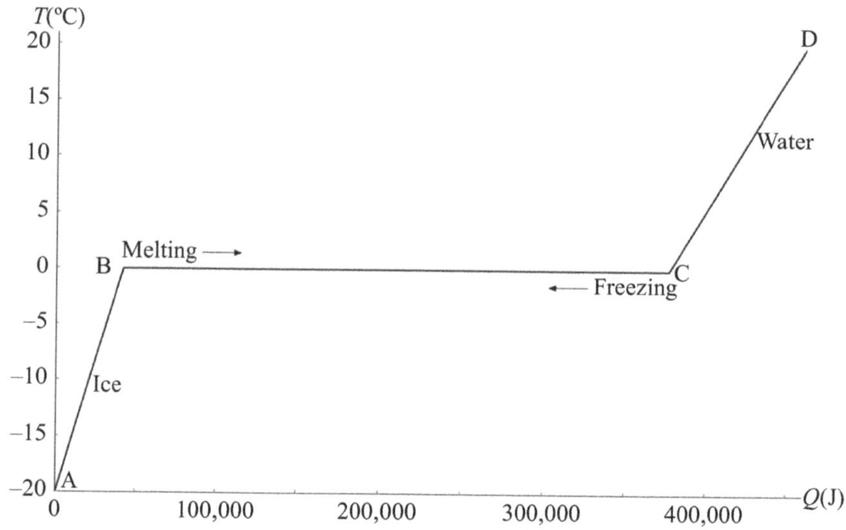


Figure 2.2 Heating ice at $-20.0\,°C$ to convert it to water at $20.0\,°C$.

The next step in the process, from B to C, is where the specific latent heat plays a role. First, notice that there is *no change in temperature* during this step. Instead, the added heat results in molecules that were part of the ice crystal being broken loose and liberated, so that they can move freely about in the liquid phase. This clearly increases the disorder and entropy of the system. The temperature isn't affected at all, however, because the ice and water are always in equilibrium during the process, and equilibrium occurs at $0\,°C$.

To find the amount of heat required to convert all of the ice to water – that is, to go from point B to point C in Figure 2.2 – we use the *specific latent heat of fusion, $l_f$, for water*, as given in Table 2.2. This yields

$$Q_{BC} = ml_f = (1.0\ \text{kg})(335{,}000\ \text{J/kg}) = 335{,}000\ \text{J}.$$

After this heat is added, the system is completely liquid water, and is still at 0 °C.

Finally, we heat the liquid water until it reaches a temperature of 20.0 °C. This corresponds to the line connecting points C and D in Figure 2.2. To find the needed heat, we use the specific heat of liquid water, as given in Table 2.1. The result is

$$Q_{CD} = mc_{\text{water}}\Delta T = (1.0\ \text{kg})[4186\ \text{J/(kg}\cdot{}^\circ\text{C)}](20.0\,^\circ\text{C}) = 83{,}720\ \text{J}.$$

Adding the heats for the three processes gives the total heat $Q_{AD}$ required for the process:

$$Q_{AD} = Q_{AB} + Q_{BC} + Q_{CD} = 41{,}800\ \text{J} + 335{,}000\ \text{J} + 83{,}720\ \text{J} = 460{,}520\ \text{J}.$$

Notice that the specific latent heat is responsible for the vast majority of the heat needed for this process – clearly, a lot of heat is needed to break the bonds holding the ice molecules in place.

Let's consider a similar situation, but this time involving the specific latent heat of vaporization (Example 2.1). To be precise, we consider the process shown in Figure 2.3, where we heat 1.00 kg of water, initially at 70.0 °C, to steam at 130 °C.

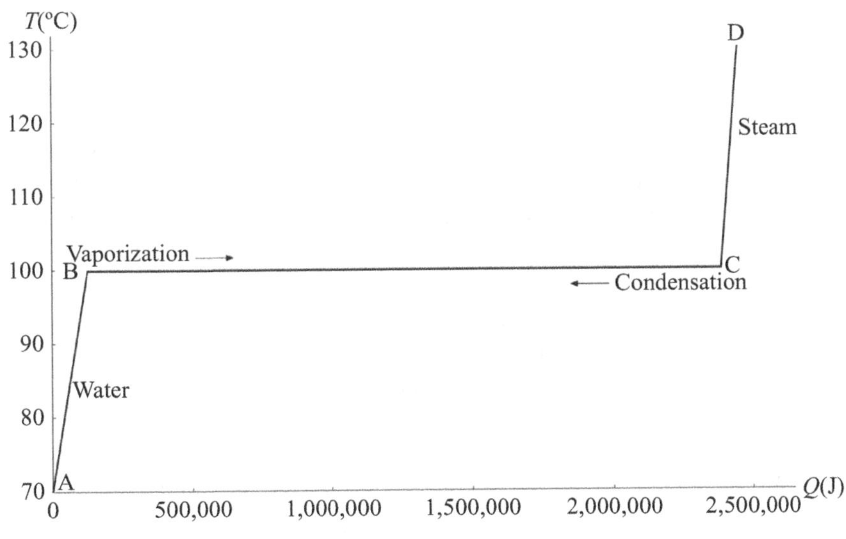


Figure 2.3 Heating water at 70.0 °C to steam at 130.0 °C.

Example 2.1 **Find the heat**

How much heat is required to raise the temperature of 1.00 kg of water at 70.0 °C to steam at 130.0 °C?

1. Find the heat required to warm the water from 70.0 °C to 100 °C. This corresponds to going from point A to point B in Figure 2.3.

$$Q_{AB} = mc_{water}\Delta T = (1.0\ \text{kg})[4186\ \text{J}/(\text{kg}\cdot{}^\circ\text{C})](30.0\,^\circ\text{C}) = 125{,}580\ \text{J}.$$

2. Next, find the heat required to boil the water at 100 °C, using the specific latent heat of vaporization, $l_v$, from Table 2.2. This corresponds to going from point B to point C in Figure 2.3. There is no change in temperature during this process.

$$Q_{BC} = ml_v = (1.0\ \text{kg})(2{,}260{,}000\ \text{J/kg}) = 2{,}260{,}000\ \text{J}.$$

3. Find the heat required to increase the temperature of the steam from 100 °C to 130.0 °C. This corresponds to going from point C to point D in Figure 2.3:

$$Q_{CD} = mc_{steam}\Delta T = (1.0\ \text{kg})[2010\ \text{J}/(\text{kg}\cdot{}^\circ\text{C})](30.0\,^\circ\text{C}) = 60{,}300\ \text{J}.$$

4. Sum these results to find the heat for the entire process from A to D:

$$Q_{AD} = Q_{AB} + Q_{BC} + Q_{CD} = 125{,}580\text{J} + 2{,}260{,}000\text{J} + 60{,}300\text{J} = 2{,}445{,}880\text{J}.$$

The majority of the heat required is due to the specific latent heat of vaporization.

To put these results in context, we present both the melting and the vaporization phase transitions we've just considered in Figure 2.4. Notice that the specific latent heat for the melting transition is considerably less than the specific latent heat for the vaporization transition. This makes sense because the change in density between the ice phase and the liquid phase is much less than it is for the liquid phase versus the steam phase. It follows that more heat is required to produce the change associated with the vaporization phase transition because additional energy is required to expand the steam phase against the surrounding atmosphere.

In Example 2.2 we use the specific latent heat to calculate the final state of the system.

Notice that the final temperature of the system just considered is 100 °C for any amount of added heat between $Q_{AB} = 125{,}580$ J and $Q_{AB} + Q_{BC} = 2{,}385{,}580$ J. Over this range of heat, the amount of liquid water will go

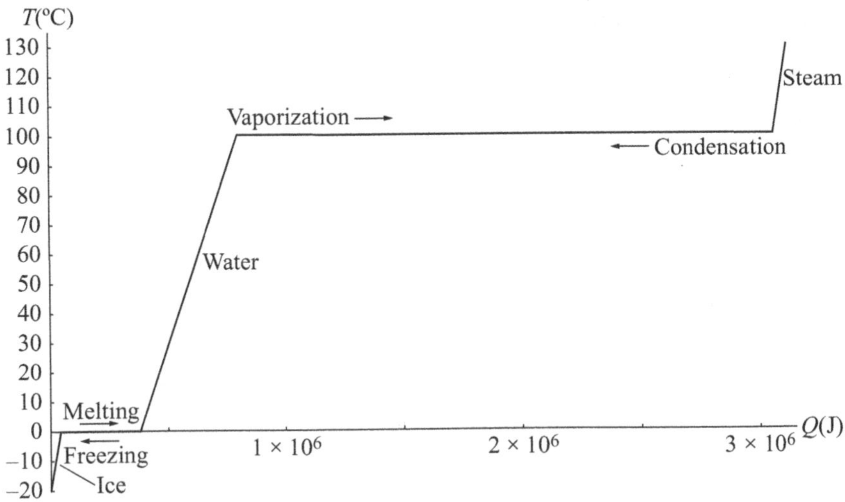


Figure 2.4 The heat involved in raising the temperature of water from −20.0 °C to 130 °C.

from 1.00 kg to 0.00 kg, and the amount of steam will go from 0.00 kg to 1.00 kg.

## Enthalpy

In various textbooks, one may find the specific latent heat expressed in terms of a change in *enthalpy*, as opposed to the heat added or removed from a system. The difference is that enthalpy refers to a process at *constant pressure*, as in a vessel open to the atmosphere, whereas heat refers to any process at all.

To begin, the enthalpy, $H$, is defined as follows:

$$H = E + PV.$$

Referring to Chapter 1, we see that the enthalpy is basically a Legendre transformation of the energy, $E$. Taking the differential of the enthalpy yields

$$dH = dE + d(PV) = (TdS - PdV) + (PdV + VdP).$$

Canceling the $PdV$ terms yields

$$dH = TdS + VdP.$$

Thus, $H = H(S, P)$ is a function of $S$ and $P$.

Example 2.2 **Find the final state**

If 2,000,000 J of heat is added to 1.00 kg of water at 70.0 °C, what is the final temperature of the system?

1. The heat required to warm the water from 70.0 °C to 100 °C was found in Step 1 of Example 2.1. The result is

$$Q_{AB} = mc_{water}\Delta T = (1.0 \text{ kg})[4186 \text{ J}/(\text{kg}\cdot{}^{\circ}\text{C})](30.0\,{}^{\circ}\text{C}) = 125{,}580 \text{ J}.$$

2. After raising the temperature of the water to 100 °C, how much heat is left over to boil water to steam at 100 °C?

$$Q_{boil} = 2{,}000{,}000 \text{ J} - Q_{AB} = 2{,}000{,}000 \text{ J} - 125{,}580 \text{ J} = 1{,}874{,}420 \text{ J}.$$

3. What is the mass of water vaporized to steam with this amount of heat? To find this quantity, rearrange the definition of specific latent heat, $l = \Delta Q/m$, to solve for the mass, $m$.

$$m_{steam} = \frac{Q_{boil}}{l_v} = \frac{1{,}874{,}420 \text{ J}}{2{,}260{,}000 \text{ J/kg}} = 0.829 \text{ kg}.$$

4. We see that less than 1.00 kg of the liquid water is vaporized; therefore, the final temperature of the system is 100 °C. The amount of liquid water remaining is

$$m_{water} = 1.00 \text{ kg} - 0.829 \text{ kg} = 0.171 \text{ kg}.$$

At constant pressure, it follows that $dP = 0$. In this case, the differential of the enthalpy is

$$dH = TdS \qquad \text{(constant pressure)}.$$

Recalling that $dQ = TdS$, we see that at constant pressure the differential of the enthalpy is simply the heat,

$$dH = dQ.$$

Because so many processes in the laboratory occur at constant atmospheric pressure, it follows that in these cases the enthalpy is the same as the heat. More generally, however, for processes that may not be at constant pressure, it is the heat that is important to a phase transition, not the enthalpy. Therefore, in the interests of generality, throughout the remainder of this book we will refer to heat, and not enthalpy.

## 2.4 The Clausius–Clapeyron Equation

We introduced the Clausius–Clapeyron equation in Chapter 1. To recap briefly, we focused there on the differential of the Gibbs free energy, $dG$, given by

$$dG = -SdT + VdP.$$

This relation shows that the independent variables of $G$ are $T$ and $P$, which, in turn, are the variables we use in a $P$-versus-$T$ phase diagram. Equating the Gibbs free energy for two phases, 1 and 2, that coexist on a phase boundary yields

$$\frac{dP}{dT} = \frac{(S_1 - S_2)}{(V_1 - V_2)} = \frac{\Delta S}{\Delta V}. \tag{2.6}$$

As mentioned in Chapter 1, this relation is not particularly useful in a typical experimental situation because the entropy $S$ is not readily measured.

We can address this problem by noting that the change in entropy at a phase boundary, $\Delta S$, is due to the latent heat, $ml$, that is added to the system to produce the change in phase. Specifically, the connection is

$$ml = \Delta Q = T\Delta S.$$

Solving this relation for $\Delta S$ and substituting in the Clausius–Clapeyron equation yields

$$\frac{dP}{dT} = \frac{\Delta S}{\Delta V} = \frac{ml}{T\Delta V} = \frac{ml}{T(V_{\mathrm{f}} - V_{\mathrm{i}})}. \tag{2.7}$$

All of the quantities in this expression – mass, latent heat, temperature, and volume – are readily measured. In addition, the subscript f refers to the final phase – after the heat $ml$ is added – and the subscript i refers to the initial phase – before the heat is added.

Before continuing, let's verify that the dimensions of this equation are correct. On the left-hand side the units are pressure over temperature, $P/T$. On the right-hand side, we see a $T$ in the denominator, so that checks out, but we need to look more carefully at the rest of the expression to verify that it has the dimensions of pressure. First, we note that $ml$ has the units of energy, which in turn has the units of $PV$. This quantity is divided by $V$. Thus, it follows that the rest of the terms combine to have the units of $P$, and the equation is dimensionally consistent, with the units of $P/T$ on both sides.

Another interesting aspect of this equation is that at first glance it appears that the slope of the phase boundary is proportional to the mass $m$ of the system. This, however, can't be correct – the slope of the phase boundary for 1 kg of a substance is the same as for 2 kg of the substance; it's a property of the substance itself, and is independent of mass. The resolution of this is that the volume in the denominator is also proportional to the mass, thus canceling the mass in the numerator.

We can see this by re-expressing the Clausius–Clapeyron equation in terms of the density of a substance, $\rho$, rather than the volume. Recall that the density is the mass per volume of the material:

$$\rho = \frac{m}{V}.$$

Solving for the volume gives

$$V = \frac{m}{\rho}.$$

With this substitution we find

$$\frac{dP}{dT} = \frac{ml}{T(V_f - V_i)} = \frac{ml}{T\left(\frac{m}{\rho_f} - \frac{m}{\rho_i}\right)}.$$

Clearly, the mass does indeed cancel. Rearranging yields

$$\frac{dP}{dT} = \frac{l\rho_i\rho_f}{T(\rho_i - \rho_f)}.$$

As expected, all of the quantities on both sides of Eq. (2.7) are *intensive*; that is, they are independent of the amount of material present in the system.

Let's apply this result to a couple of special cases with different types of phase boundaries.

## Solid–Liquid Phase Boundary

Consider a system consisting of a quantity of solid that is on the solid–liquid phase boundary at temperature $T$. It follows that the initial phase of the system has density $\rho_i = \rho_{solid}$. Adding heat to the system converts some of the solid to the liquid phase. Eventually, when the heat per kilogram $l_f$ has been added, the system is in the final state, a liquid with the density $\rho_f = \rho_{liquid}$. With these substitutions, the Clausius–Clapeyron equation takes the form

$$\frac{dP}{dT} = \frac{l_f \rho_{solid} \rho_{liquid}}{T\left(\rho_{solid} - \rho_{liquid}\right)}. \tag{2.8}$$

All of the quantities in this expression are readily determined experimentally.

One thing to note about Eq. (2.8) is that the name of the specific latent heat of fusion, $l_f$, in the numerator is a bit unfortunate. Why is this? Well, the term "fusion" refers to the phase transition that occurs when a liquid fuses to form a solid. Clearly, this transition occurs when heat is *removed* from a system. The specific latent heat, however, is always defined to be a *positive quantity*, which means that it refers to heat that is *added* to a system. Thus, it makes more sense to refer to the heat for this type of phase transition as the specific latent heat of melting, rather than fusion. In what follows, however, we will continue to use the inappropriate name $l_f$ for this specific latent heat, given that it is firmly entrenched in the scientific literature.

Another interesting aspect of Eq. (2.8) is the fact that the slope $dP/dT$ can be positive or negative, depending on the densities of the solid and liquid phases. For most substances, the density of the solid is greater than that of the liquid. In such cases, we see that $dP/dT$ is positive. On the other hand, in those exceptional cases where the liquid is denser than the solid – as in water – then $dP/dT$ is negative. We shall explore the implications of this is detail in Chapter 3.

For small changes in temperature and pressure, the **Clausius–Clapeyron equation for a solid–liquid boundary** can be written as

$$\Delta P = \left[ \frac{l_f \rho_{\text{solid}} \rho_{\text{liquid}}}{T\left(\rho_{\text{solid}} - \rho_{\text{liquid}}\right)} \right] \Delta T. \tag{2.9}$$

Notice that the quantity in square brackets is a constant, with the dimensions of pressure per kelvin. As an example, consider the case of the solid–liquid phase transition in gold. The density of the solid is $\rho_{\text{solid}} = 18{,}300\ \text{kg/m}^3$, the density of the liquid is $\rho_{\text{liquid}} = 17{,}190\ \text{kg/m}^3$, and the specific latent heat of fusion in gold is $l_f = 62{,}800\ \text{J/kg}$. In addition, the melting temperature is $T = 1337\ \text{K}$. Using these values, we find that the change in pressure for the melting transition is related to the change in temperature as follows:

$$\Delta P = (13.3\ \text{MPa/K})\Delta T.$$

We see that as the pressure is increased, $\Delta P > 0$, the melting temperature increases as well, $\Delta T > 0$. This makes sense physically, because as the pressure is increased, the molecules in the solid are held more firmly in place, and hence the temperature must be increased to cause the solid to melt. As a specific example, an increase in pressure of 133 atm causes an increase in the melting temperature of gold by 1.00 K.

### Integrating the Solid–Liquid Phase Boundary

To see how the pressure on the phase boundary changes over a larger range of temperature, we can integrate the solid–liquid Clausius–Clapeyron equation in Eq. (2.8). Returning to the differential limit and rearranging, we have

$$dP = \left[\frac{l_f \rho_{\text{solid}} \rho_{\text{liquid}}}{\left(\rho_{\text{solid}} - \rho_{\text{liquid}}\right)}\right] \frac{dT}{T}.$$

Assuming that the densities and specific latent heat are constant over the range of temperatures and pressures of interest, we can write

$$\int_{P_1}^{P_2} dP = \left[\frac{l_f \rho_{\text{solid}} \rho_{\text{liquid}}}{\left(\rho_{\text{solid}} - \rho_{\text{liquid}}\right)}\right] \int_{T_1}^{T_2} \frac{dT}{T}.$$

Integrating yields another expression for the Clausius–Clapeyron equation:

$$P_2 - P_1 = \left[\frac{l_f \rho_{\text{solid}} \rho_{\text{liquid}}}{\left(\rho_{\text{solid}} - \rho_{\text{liquid}}\right)}\right] \ln\left(\frac{T_2}{T_1}\right). \tag{2.10}$$

To see the connection between the integral relationship and the linear approximation in Eq. (2.9), consider a small change in temperature from $T$ to $T + \Delta T$. Substituting this into Eq. (2.10) yields

$$\Delta P = \left[\frac{l_f \rho_{\text{solid}} \rho_{\text{liquid}}}{\left(\rho_{\text{solid}} - \rho_{\text{liquid}}\right)}\right] \ln\left(\frac{T + \Delta T}{T}\right).$$

Expanding the logarithm, using $\ln(1 + x) \approx x$ for small $x = \Delta T / T$, we find

$$\Delta P = \left[\frac{l_f \rho_{\text{solid}} \rho_{\text{liquid}}}{\left(\rho_{\text{solid}} - \rho_{\text{liquid}}\right)}\right] \frac{\Delta T}{T} = \left[\frac{l_f \rho_{\text{solid}} \rho_{\text{liquid}}}{T\left(\rho_{\text{solid}} - \rho_{\text{liquid}}\right)}\right] \Delta T.$$

This agrees with our earlier linear approximation.

### Integrating the Liquid–Vapor Phase Boundary

Next, we consider the Clausius–Clapeyron equation as applied to the liquid–vapor phase boundary at temperature $T$. The initial phase is the liquid, and hence $\rho_i = \rho_{\text{liquid}}$. Adding the specific latent heat of vaporization per kilogram,

$l_v$, converts the system to the gaseous phase, with the density $\rho_f = \rho_{\text{vapor}}$. It follows that

$$\frac{dP}{dT} = \frac{l_v \rho_{\text{liquid}} \rho_{\text{vapor}}}{T\left(\rho_{\text{liquid}} - \rho_{\text{vapor}}\right)}. \tag{2.11}$$

As before, all of the quantities in this equation are intensive and readily measurable. Furthermore, in all common cases the density of the liquid is greater than the density of the vapor, and thus the slope of the liquid–vapor phase boundary is positive.

This equation is fine as it is, but there are two useful approximations that give additional insight into the physics of the liquid–vapor phase boundary. We consider these approximations next, both of which involve specific properties of the vapor phase.

**Approximation 1: The Vapor Is Much Less Dense Than the Solid**

The first approximation comes from the observation that the density of the vapor phase is generally *much less* than that of the liquid phase. In fact, the density of the vapor phase is often so small that we can ignore it completely in comparison with that of the solid. For example, the vapor phase of water – that is, steam – is roughly 1600 times less dense than the liquid phase. With this in mind, we can make the following replacement in the denominator of Eq. (2.11):

$$\rho_{\text{liquid}} - \rho_{\text{vapor}} \approx \rho_{\text{liquid}}.$$

Thus, the Clausius–Clapeyron equation can be written as follows:

$$\frac{dP}{dT} = \frac{l_v \rho_{\text{liquid}} \rho_{\text{vapor}}}{T\left(\rho_{\text{liquid}} - \rho_{\text{vapor}}\right)} \approx \frac{l_v \rho_{\text{liquid}} \rho_{\text{vapor}}}{T \rho_{\text{liquid}}} = \frac{l_v \rho_{\text{vapor}}}{T}. \tag{2.12}$$

It's interesting to note – if not a bit surprising, in fact – that even though the density of the vapor phase is negligible compared to that of the solid, it is the vapor density that remains in the final form of the equation and is responsible for determining the slope of the phase boundary.

We can get additional insight into the importance of the vapor phase by returning to the form of the Clausius–Clapeyron equation given in Eq. (2.7). There, we see that

$$\frac{dP}{dT} = \frac{ml}{T(V_f - V_i)}.$$

When applied to the liquid-to-vapor phase transition, we have $V_f = V_{\text{vapor}}$, $V_i = V_{\text{liquid}}$, and $l = l_v$. Returning to the observation that the density of the vapor can be as much as 1600 times *less* than that of the liquid, it follows that the volume of the vapor can be 1600 times *greater* than that of the liquid for a given amount of mass. Thus, it is appropriate to ignore the volume of the liquid compared to that of the vapor. That is,

$$V_{\text{vapor}} - V_{\text{liquid}} \approx V_{\text{vapor.}}$$

With this substitution we have

$$\frac{dP}{dT} = \frac{ml_{\text{v}}}{TV_{\text{vapor}}}.$$

Finally, noting that $\rho_{\text{vapor}} = m/V_{\text{vapor}}$, we recover Eq. (2.12). Thus, we see that the reason the vapor pressure is most important in the Clausius–Clapeyron equation is that the liquid has an insignificant volume.

**Approximation 2: The Vapor Is Approximately an Ideal Gas**

The second approximation we make is to consider the low-density vapor as an ideal gas. Recall that the equation of state for an ideal gas can be written as

$$PV = Nk_{\text{B}}T.$$

Applying this to the vapor phase, we can write the volume of the vapor in the following form:

$$V_{\text{vapor}} = \frac{Nk_{\text{B}}T}{P}.$$

We substitute this result into the Clausius–Clapeyron equation to find

$$\frac{dP}{dT} = \frac{ml_{\text{v}}}{T\left(\frac{Nk_{\text{B}}T}{P}\right)}.$$

For the mass $m$ of the gas, note that we can express it as the mass of an individual molecule, $\mu$, times the number of molecules in the system, $N$. That is,

$$m = \mu N.$$

Substituting this in the Clausius–Clapeyron equation yields

$$\frac{dP}{dT} = \frac{\mu N l_{\text{v}}}{T\left(\frac{Nk_{\text{B}}T}{P}\right)} = \frac{\mu l_{\text{v}} P}{k_{\text{B}}T^2}.$$

As expected, the number of molecules, $N$, has canceled – the slope of the phase boundary is independent of the number of molecules in the system.

Rearranging this equation, and integrating from an initial temperature and pressure $T_1$and $P_1$, to a final temperature and pressure $T_2$ and $P_2$, we find

$$\int_{P_1}^{P_2} \frac{dP}{P} = \left(\frac{\mu l_v}{k_B}\right) \int_{T_1}^{T_2} \frac{dT}{T^2}.$$

Carrying out the integrals on both sides of the equation yields

$$\ln P_2 - \ln P_1 = -\left(\frac{\mu l_v}{k_B}\right)\left(\frac{1}{T_2} - \frac{1}{T_1}\right).$$

Thus, the *Clausius–Clapeyron equation for a liquid–vapor phase boundary* can be written as follows:

$$P_2 = P_1 e^{-\left(\frac{\mu l_v}{k_B}\right)\left(\frac{1}{T_2} - \frac{1}{T_1}\right)}. \tag{2.13}$$

In this equation, note that the temperature is measured in kelvins, but the pressure can be measured in any desired units.

As a specific example, let's consider the case of the vaporization of water. We will make the following identifications: $P_1 = 1.00$ atm, $T_1 = 100\,°\text{C} = 373.15$ K, $P_2 = P$, and $T_2 = T$. In addition, given that the mass of a water molecule is $\mu = 2.99 \times 10^{-26}$ kg, we have

$$\frac{\mu l_v}{k_B} = \frac{(2.99 \times 10^{-26}\ \text{kg})(2{,}260{,}000\ \text{J/kg})}{1.38 \times 10^{-23}\ \text{J/K}} = 4897\ \text{K}.$$

Combining these results, Eq. (2.13) can be written as follows:

$$P = (1.00\ \text{atm})e^{-(4897\ \text{K})\left(\frac{1}{T} - \frac{1}{373.15\ \text{K}}\right)}. \tag{2.14}$$

We plot this function in Figure 2.5. The dashed lines in the plot indicate the boiling temperature at atmospheric pressure.

Note that the boiling temperature increases with pressure. This is as expected, because the greater the pressure applied to a liquid the more energy it takes to release a molecule from the liquid into the vapor. The additional energy is supplied by an increase in temperature of the boiling liquid.

## Using Molar Quantities

In thermodynamic literature, it's common to see the Clausius–Clapeyron equation expressed in terms of molar quantities, as opposed to quantities that are given on a per molecule basis. The conversion between molecular and

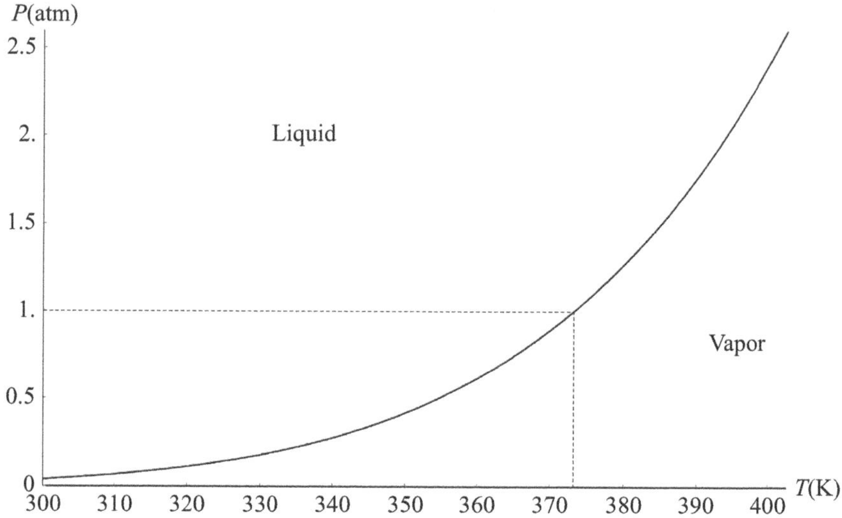


Figure 2.5 Phase boundary for the liquid–vapor phase transition in water, as given in Eq. (2.14). At one atmosphere, the boiling temperature is 273.15 K + 100 K = 373.15 K, as indicated by the dashed lines.

molar quantities is straightforward, but it's useful to go through each step of the process carefully to make it completely clear.

We start with the quantity in Eq. (2.13) that needs to be converted:

$$\frac{\mu l_{\mathrm{v}}}{k_{\mathrm{B}}}.$$

Note that $\mu$ is the mass *per molecule*, and that $k_{\mathrm{B}}$ is also defined on a *per-particle* basis. We now re-express this quotient in terms of quantities that are measured *per mole*.

First, we multiply and divide this quantity by the number of particles in the system, $N$, to obtain

$$\frac{\mu l_{\mathrm{v}}}{k_{\mathrm{B}}} = \frac{N\mu l_{\mathrm{v}}}{N k_{\mathrm{B}}}.$$

In the numerator, the quantity $N\mu$ is equal to the mass of the system, $m$. In the denominator, we note that $Nk_{\mathrm{B}}$ is equal to $nR$, where $n$ is the number of moles, and $R$ is the universal gas constant, $R = 8.314\ \mathrm{J/K \cdot mol}$. Therefore,

$$\frac{\mu l_{\mathrm{v}}}{k_{\mathrm{B}}} = \frac{m l_{\mathrm{v}}}{nR}.$$

We define the molar mass, $M_{\mathrm{m}}$, to be the mass per mole; that is, $M_{\mathrm{m}} = m/n$. Thus,

$$\frac{\mu l_v}{k_B} = \frac{M_m l_v}{R}.$$

Finally, we define $M_m l_v$ to be the molar latent heat of vaporization, $l_{mv}$:

$$l_{mv} = M_m l_v$$

As a result, the quotient of interest can be written as

$$\frac{\mu l_v}{k_B} = \frac{l_{mv}}{R}$$

Note that the units of $l_{mv}$ are J/mol, and the units of $l_{mv}/R$ are K, as expected. This gives

$$P_2 = P_1 e^{-\left(\frac{l_{mv}}{R}\right)\left(\frac{1}{T_2} - \frac{1}{T_1}\right)}.$$

As before, the pressure can be measured in any units, as long as they are consistent on the two sides of the equation.

Let's take a look at a specific example. For water, the molar mass is due to two hydrogen atoms (molar mass 1 for each) plus one oxygen atom (molar mass 16). Including the effects of isotopes with different masses, the molar mass of water is $M_m = 18.015\ \mathrm{g/mol}$. Expressing the specific latent heat of vaporization for water in units of J/g, to cancel grams in the molar mass, we have

$$l_{mv} = (18.015\ \mathrm{g/mol})(2260\ \mathrm{J/g}) = 40{,}714\ \mathrm{J/mol}.$$

Finally, the quotient of interest is

$$\frac{\mu l_v}{k_B} = \frac{l_{mv}}{R} = \frac{40{,}714\ \mathrm{J/mol}}{8.314\ \mathrm{J/K \cdot mol}} = 4897\ \mathrm{K}.$$

Thus, we see that with molar quantities we have precisely the same result for the pressure on the phase boundary given in Eq. (2.14):

$$P = (1.00\ \mathrm{atm})e^{-(4897\ \mathrm{K})\left(\frac{1}{T} - \frac{1}{373.15\ \mathrm{K}}\right)}.$$

For convenience, we present molar latent heats of fusion and vaporization, with the units J/mol, for a variety of substances in Table 2.3.

## 2.5 Heat Engines and Refrigerators

In this section, we consider some of the practical applications of heat. As we shall see, it's possible to tap into the heat energy that is transferred from a hot

Table 2.3 *Molar latent heats of fusion and vaporization for various substances*

| Substance | Specific latent heat of fusion, $L_{mf}$ (J/mol) | Specific latent heat of vaporization, $L_{mv}$ (J/mol) |
|---|---|---|
| Aluminum | 8660 | 307,600 |
| Benzene | 10,000 | 30,500 |
| Copper | 13,200 | 322,100 |
| Gold | 13,200 | 310,900 |
| Iron | 11,700 | 354,100 |
| Lead | 4640 | 180,500 |
| Nitrogen | 715 | 5600 |
| Sodium | 2600 | 97,420 |
| Water | 6020 | 40,700 |

object to a cold object and use a portion of it to run an engine. This can be done without a change in phase, but using the latent heat of a phase transition can greatly enhance the utility of such engines. We will explore applications of latent heat to both steam engines and refrigerators in this section.

## Heat Engines

One statement of the second law of thermodynamics is as follows: *Heat energy is transferred spontaneously from a hot object to a cold object when they are brought into contact.* We illustrate this in Figure 2.6. Here we see a region at high temperature, denoted $T_h$ for $T$ hot, connected to another region at a lower temperature, indicated by $T_c$ for $T$ cold. When these regions are connected, a quantity of heat energy $|\Delta Q|$ transfers out of the high-temperature region and into the low-temperature region.

Assuming the transfer of energy occurs reversibly, we can relate the heat transfer $\Delta Q$ to the change in entropy $\Delta S$ as follows:

$$\Delta Q = T\Delta S.$$

It follows that the *total* change in entropy for the system due to this transfer of heat is given by

$$\Delta S_{\text{total}} = \Delta S_h + \Delta S_c = -\frac{|\Delta Q|}{T_h} + \frac{|\Delta Q|}{T_c}.$$

Notice that the entropy of the hot region decreases, $\Delta S_h < 0$, and the entropy of the cold region increases, $\Delta S_c > 0$. The quantity of heat $|\Delta Q|$ is the same in the numerator of each of these terms, but the cold region has the smaller denominator, since $T_c < T_h$. It follows that the increase of entropy in the cold region is

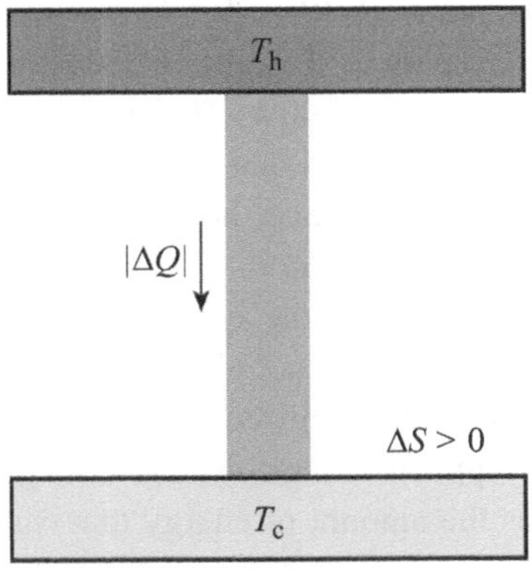


Figure 2.6 When a hot region (h) is thermally connected to a cold region (c), the heat energy $|\Delta Q|$ flows from the hot region to the cold region. During this process the entropy of the system increases, $\Delta S > 0$.

greater than the decrease of entropy in the hot region, and hence the total change in entropy is positive, $\Delta S_{total} > 0$. This leads to an alternate statement of the second law of thermodynamics: *The total change in entropy of a spontaneous process is either positive or zero, but it can never be negative.*

Now, the basic idea of a heat engine is to tap into the transfer of heat energy $|\Delta Q|$ and convert some of it to mechanical work, $W$. Ideally, it would be nice if *all* of the heat energy could be converted to work, as illustrated in Figure 2.7. In this case, the heat $|Q_h|$ comes out of the hot region and goes into the heat

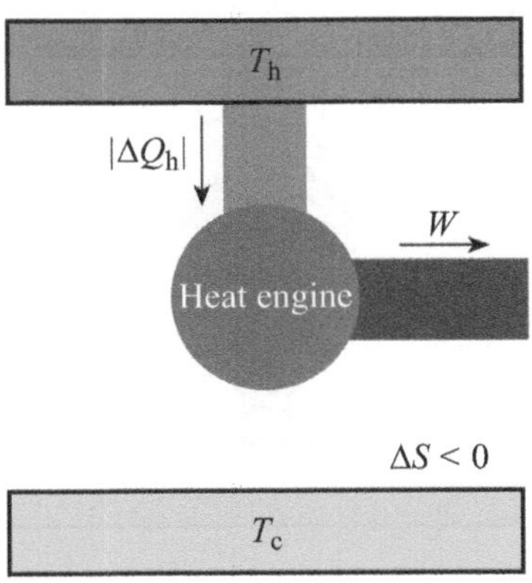


Figure 2.7 It is impossible for heat energy to flow from a hot region and be converted completely into work. Such a process would result in a decrease in the entropy of the universe, which is forbidden by the second law of thermodynamics.

engine, which produces the work $W = |Q_\mathrm{h}|$. We would say that an engine like this is 100% efficient, since all of the input heat energy appears as work.

Unfortunately, an engine with 100% efficiency is forbidden by the second law of thermodynamics, which says that the change in entropy for a case like this must be positive or zero. Notice, however, that the entropy of the hot region decreases by the amount $\Delta S_\mathrm{h} = -|Q_\mathrm{h}|/T_\mathrm{h}$, and there is no compensating increase in the entropy of the cold region, $\Delta S_\mathrm{c} = 0$. Thus, the total entropy decreases, which is simply not allowed.

To make the change in total entropy equal to or greater than zero, there must be some *exhaust heat* supplied to the cold region to give a positive change in entropy. This decreases the amount of energy that can be converted to work, but it produces the necessary increase in entropy to offset the decrease in entropy of the hot region. This is illustrated in Figure 2.8.

Now, the *greatest* amount of work is produced when the amount of exhaust heat is the *least* it can be, which occurs when the increase in entropy of the cold region is just enough to make the total change in entropy equal to zero. We can express this as follows:

$$\Delta S_\mathrm{total} = \Delta S_\mathrm{h} + \Delta S_\mathrm{c} = -\frac{|\Delta Q_\mathrm{h}|}{T_\mathrm{h}} + \frac{|\Delta Q_\mathrm{c}|}{T_\mathrm{c}} = 0.$$

Rearranging yields

$$\frac{|\Delta Q_\mathrm{h}|}{T_\mathrm{h}} = \frac{|\Delta Q_\mathrm{c}|}{T_\mathrm{c}}.$$

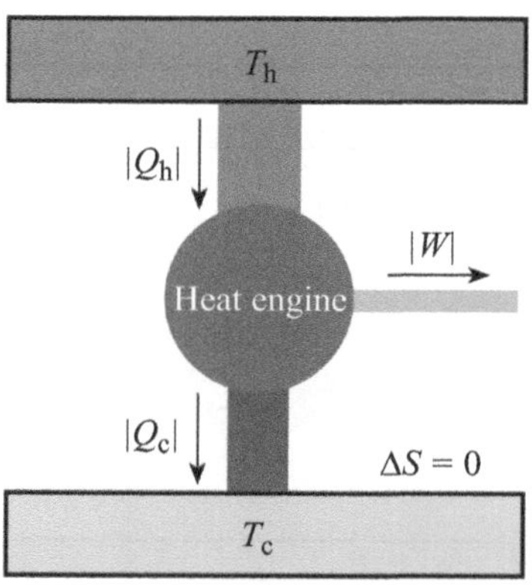


Figure 2.8 In a real heat engine, a minimum amount of exhaust heat $|\Delta Q_\mathrm{c}|$ is required for the entropy change of the system to be zero, $\Delta S = 0$. The amounts of heats and work in this figure, as measured by the thickness of the lines entering and leaving the heat engine, correspond to an efficiency of $e = 0.268$.

Finally,

$$\frac{|\Delta Q_c|}{|\Delta Q_h|} = \frac{T_c}{T_h}.$$

Notice that the minimum amount of exhaust heat is directly related to the absolute temperatures of the hot and cold regions.

The work done by the heat engine is given by energy conservation. The energy $|\Delta Q_h|$ goes into the engine, the amount $|\Delta Q_c|$ is expelled, and hence the work, $W$, done by the engine is

$$W = |\Delta Q_h| - |\Delta Q_c|.$$

Now, we define the efficiency of the heat engine, $e$, to be the fraction of heat supplied to the engine that appears as useful work. Thus,

$$e = \frac{W}{|\Delta Q_h|} = \frac{|\Delta Q_h| - |\Delta Q_c|}{|\Delta Q_h|} = 1 - \frac{|\Delta Q_c|}{|\Delta Q_h|} = 1 - \frac{T_c}{T_h}.$$

This is the *maximum efficiency* of a heat engine operating between the absolute temperatures $T_h$ and $T_c$; it is a fundamental limit imposed by the second law of thermodynamics.

As a specific example, suppose we have a heat engine operating between the temperatures of boiling water, at 100 °C = 373.15 K, and freezing water, at 0 °C = 273.15 K. The maximum possible efficiency of this heat engine is

$$e = 1 - \frac{T_c}{T_h} = 1 - \frac{273.15\ \text{K}}{373.15\ \text{K}} = 0.268.$$

Thus, the engine has an efficiency of 26.8%. This efficiency is illustrated in Figure 2.8, where the work coming out of the heat engine is 26.8% of the energy supplied to the engine at the hot temperature. The remaining 73.2% of the energy is exhausted at the cold temperature. No heat engine of any possible design operating between these temperatures can have a greater efficiency.

## Carnot Cycle

An explicit representation of an ideal heat engine on a $P$–$V$ diagram is referred to as a *Carnot cycle*, named for the French physicist Sadi Carnot who presented it in 1824. We show a sample Carnot cycle in Figure 2.9.

Notice that the Carnot cycle consists of four separate processes linked together to form a closed circuit. We describe these processes for the case of a heat engine consisting of a quantity of gas in a cylinder with a moveable piston connected to a crankshaft:

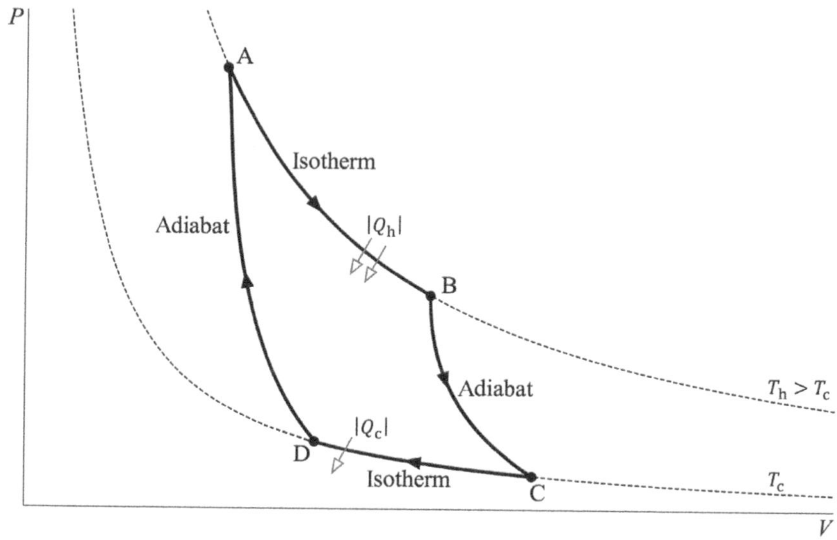


Figure 2.9 The Carnot cycle on a $P$–$V$ plot. The cycle consists of two isothermal processes and two adiabatic processes. The efficiency of this cycle is given by $e = 1 - T_c/T_h$.

**(A – B) Isothermal expansion:** The cylinder is in thermal contact with a high-temperature bath that is held at the constant temperature $T_h$. The piston does work on the external world as the gas expands. As the volume of the gas increases, its pressure decreases.

**(B – C) Adiabatic expansion:** The cylinder is now insulated so that no heat can flow through its walls. Thus, the expansion of the gas from B to C is referred to as an *adiabatic* (a-dia-batic = no-flow through) *process*. As the volume of the gas increases, the pressure drops even more rapidly than it does in the isothermal expansion. The heat engine continues to do work on the external world during this process.

**(C – D) Isothermal compression:** Work is done *on* the engine during the isothermal compression from C to D. The cylinder is in thermal contact with a low-temperature bath, and the temperature is held constant at the value $T_c$. The pressure of the gas rises slightly as the volume decreases.

**(D – A) Adiabatic compression:** The cylinder is once again insulated so that no heat flows through its walls. Work is again done on the system during this process, compressing the gas in the cylinder and increasing its pressure sharply back to the original value at the point A.

The net effect of the cycle is that work is done on the external world in the amount $|W|$. In addition, a quantity of heat $|\Delta Q_h|$ is supplied to the engine, and a quantity of heat $|\Delta Q_c|$ is exhausted from the engine. The work done by the engine is equal to the area enclosed by the cycle and, by energy conservation, has the value $|W| = |\Delta Q_h| - |\Delta Q_c|$.

During the isothermal expansion, A – B, heat is drawn from the heat reservoir to maintain the constant temperature $T_h$ in the cylinder. This causes a decrease in the entropy of the environment of the amount

$$\Delta S_{AB} = -\frac{|\Delta Q_h|}{T_h}.$$

There is no entropy change during the adiabatic expansion, from B to C, since there is no heat flow. Thus,

$$\Delta S_{BC} = 0.$$

For the isothermal compression, from C to D, heat flows out of the engine into the heat bath. This causes the entropy of the environment to increase by the amount

$$\Delta S_{CD} = \frac{|\Delta Q_c|}{T_c}.$$

Finally, on the reversible diabatic compression, from D to A, there is no heat flow, and hence no change in entropy:

$$\Delta S_{DA} = 0.$$

The total change in entropy of the system is

$$\Delta S_{total} = \Delta S_{AB} + \Delta S_{BC} + \Delta S_{CD} + \Delta S_{DA} = -\frac{|\Delta Q_h|}{T_h} + 0 + \frac{|\Delta Q_c|}{T_c} + 0.$$

The best situation is no entropy change for the cycle, and hence

$$\frac{|\Delta Q_h|}{T_h} = \frac{|\Delta Q_c|}{T_c}.$$

This is the same condition presented earlier in this section, and leads to the same efficiency for the Carnot cycle heat engine, $e = 1 - T_c/T_h$.

## Steam Engines

Heat engines come in a variety of styles and sizes, ranging from the relatively small internal combustion engines in automobiles and motorcycles, to the massive engines on jet airplanes, and the truly immense engines that power rockets on liftoff. There is one type of heat engine, however, that stands out

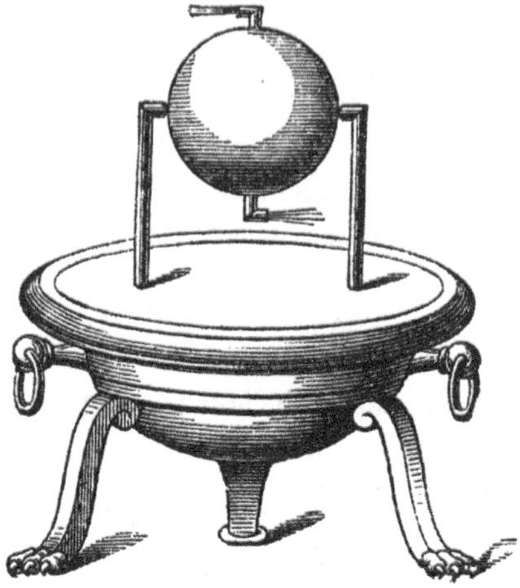

Figure 2.10 The aeolipile of Hero, an early illustration of steam power. Source: Nastasic/DigitalVision Vectors/Getty Images.

from all the rest, and plays a preeminent role in the development of modern society: the steam engine.

Steam engines are powered by the fact that when water vaporizes to steam it expands in volume by about 1600 times. This incredible expansion can be tapped to produce mechanical work. The first demonstration of the power of steam is credited to Hero of Alexandria (circa AD 62) with his invention of the aeolipile, shown in Figure 2.10. In this device, heat supplied to a vessel containing water produces steam, which expands through pipes into a moveable sphere with two oppositely directed nozzles attached to it. As the steam exits through the nozzles, it causes the sphere to rotate rapidly about its axle.

Hero's aeolipile wasn't connected to any machinery to produce useful work – it was simply a device to illustrate the power of steam in a dramatic fashion. It took another 1600 years before the expansion of steam was harnessed to practical effect by the English engineer Thomas Savery, who patented a crude oscillatory-motion steam engine in 1698, and used it to pump water out of coal mines. In this way, we can say that the steam engine ushered in the Industrial Revolution that has shaped modern civilization.

In 1765, the Scottish engineer James Watt invented the first steam engine with a modern-style rotary motion. This greatly enhanced the utility of steam engines, which soon after began to appear not only in coal mines, but in factories, locomotives, ships, and even cars. Watt's steam engine was only about 2% to 3% efficient – certainly a small efficiency, but much greater than that of previous steam engines, and still enough to do an enormous amount of useful work in practical situations. The efficiency of steam engines improved considerably over time until, in the early twentieth century, the efficiencies reached 20% to 25%.

The operation of a steam engine is shown in simplified form in Figure 2.11. First, in Figure 2.11(a), we see a furnace burning a fuel – such as coal for example – and using that heat to boil liquid water and convert it to steam. This occurs at a temperature of 100 °C.

Note that the temperature in the boiling chamber remains at 100 °C as the latent heat of vaporization is added to the water, producing more and more

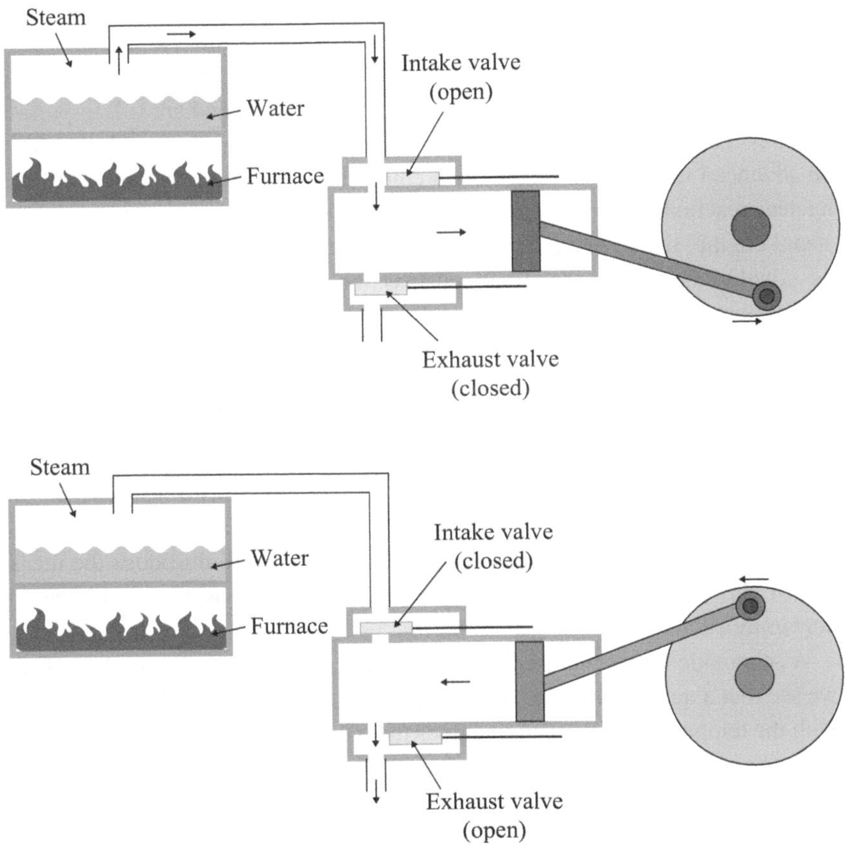


Figure 2.11 The operation of a steam engine. (a) The intake valve is open, and steam expands into the cylinder, pushing the piston to the right. (b) The rotary motion of the axle returns the piston to the left as the expanded steam exits freely through the exhaust valve.

steam. Thus, the added heat energy goes into an expansion of liquid water to steam rather than an increase in temperature – the advantage of this is that if the heat simply increased the temperature, it would rise to values so large that the engine would be destroyed.

The expanding steam created in the boiling chamber enters the cylinder of the steam engine through the open intake value, pushing the piston to the right. Notice that the exhaust valve is closed at this point of the engine's cycle. The piston is attached to a crankshaft that rotates an axle and does mechanical work on the external world.

Once the piston has moved as far as possible to the right, the intake valve closes and the exhaust valve opens, as we see in Figure 2.11(b). The continued rotary motion of the crankshaft pushes the piston back to the left, which allows the expanded steam to exit out of the exhaust valve into the air, or into a chamber where the steam is condensed back to liquid form. In a very real sense, steam engines powered the Industrial Revolution, and are still important in today's society. For example, most nuclear power plants, though they rely on advanced technology, ultimately are heat engines. They generate heat from nuclear reactions to boil water and produce steam. The steam then powers a steam engine, which generates electricity. So even with today's high technology, including nuclear reactions and mass being converted to energy, steam engines are still an important part of the modern world.

## Refrigerators

Now that we've explored a number of different heat engines, we turn to refrigerators, which, as we shall see, can be considered as a heat engine run in reverse. In what follows, we'll consider the most common type of refrigerator, the vapor-compression refrigerator, and we'll introduce the measure of the effectiveness of a refrigerator, referred to as the *coefficient of performance*.

A schematic representation of a refrigerator is shown in Figure 2.12. Here we see that a quantity of heat, $|Q_c|$, is removed from a low-temperature region, with the temperature $T_c$, an engine does the work $|W|$, and the quantity of heat, $|Q_h|$, is exhausted to a high-temperature region, with the temperature $T_h$. The low-temperature region can be thought of as the interior of a refrigerator, which we want to keep cool, and the high-temperature region is the room temperature of a kitchen.

Notice how this contrasts with a heat engine, which taps into the natural flow of heat energy from high temperature to low temperature to do mechanical work. In a refrigerator, heat energy flows in the *opposite direction* – from

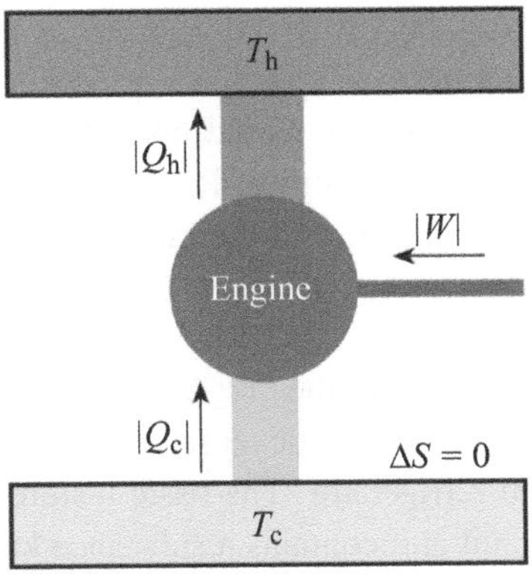


Figure 2.12 A schematic illustration of a refrigerator with a coefficient of performance equal to 4, as indicated by the thickness of the lines entering and leaving the engine. The refrigerator takes the heat $|Q_c|$ from the cold region, does the work $|W|$, and delivers the heat $|Q_h|$ to the hot region.

the cold region to the warm region. It's a bit like comparing the water pouring over a waterfall, which is the natural direction of flow, to water being *pumped upward* from the bottom of the falls to the top, which requires mechanical work to be done. The engine in the refrigerator shown in Figure 2.12 is doing the mechanical work $|W|$ to "pump" the heat energy in the opposite direction, from cold to hot.

The effectiveness of a refrigerator is measured by the *coefficient of performance, COP,* defined as follows:

$$\text{COP} = \text{coefficient of performance} = \frac{|Q_c|}{|W|}.$$

Notice that the COP measures the amount of heat removed from the cold region in terms of the work required to operate the refrigerator. Using energy conservation, we can write the following condition that applies to the work and heat:

$$|Q_c| + |W| = |Q_h|.$$

Solving for the work we have

$$|W| = |Q_h| - |Q_c|.$$

Thus, it follows that

$$\text{COP} = \frac{|Q_c|}{|Q_h| - |Q_c|}.$$

Notice that the COP can be expressed entirely in terms of heats, with no reference to the temperature.

In the ideal case, a refrigerator would operate with no work being done, which would give a COP of infinity. Real-world refrigerators have much smaller COPs, of course, typically in the range of 3 to 5. Figure 2.12 shows the case for a refrigerator with COP $= 4$. In this case, a work of $|W| = 25$ J takes the heat $|Q_c| = \mathrm{COP}|W| = 100$ J from the cold region and exhausts the heat $|Q_h| = |Q_c| + |W| = 125$ J to the "high-temperature" kitchen.

In Figure 2.13 we show a schematic representation of the operating elements of a vapor-compression refrigerator. The basic feature of this device is a pipe that forms a closed circuit and contains a substance known as the *refrigerant.* The refrigerant is in the vapor phase at some points in the circuit, and is liquid at other points. The phase transitions between liquid and vapor are used to remove the heat energy $|Q_c|$ from the interior of the refrigerator, as we see at

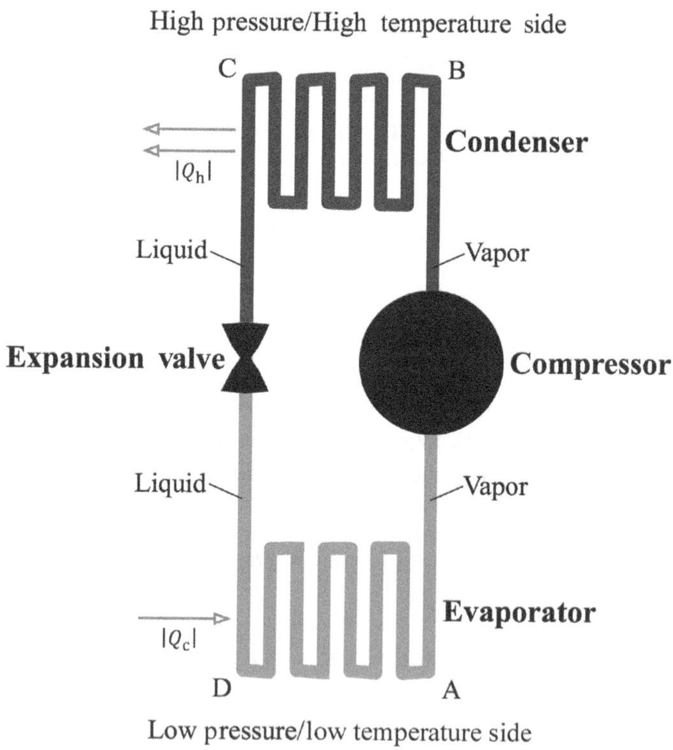


Figure 2.13 A schematic representation of the working elements of a vapor-compression refrigerator.

the bottom of the figure, and exhaust the heat energy $|Q_h|$ into the kitchen, as shown at the top of the figure.

To explore the refrigeration cycle in detail, we start at the lower right corner of the diagram, labeled A. At this point the refrigerant is a low-pressure, low-temperature vapor as it emerges from inside the refrigerator. The refrigerant then enters an engine, called the *compressor*, which compresses the vapor to a high pressure and a correspondingly high temperature, at point B. This vapor is sent to *condenser* coils, from B to C, on the outside of the refrigerator, where heat energy from the hot vapor is radiated into the cooler kitchen, causing the vapor to undergo a phase transition and condense to a high-pressure, high-temperature liquid. A significant amount of heat energy, the latent heat of vaporization, is given off during this part of the cycle, and the temperature remains constant at its high value.

Next, the pipe conducts the liquid refrigerant from C to D, where along the way it encounters a device known as the *expansion valve*. When passing through this valve the liquid expands rapidly and cools by a significant amount in an irreversible process known as "throttling."

You can experience this cooling effect yourself by breathing on your outstretched hand. If you breathe with your mouth wide open, you feel warm air on your hand. If you breathe with pursed lips, so the air expands as it exits your mouth, you feel cool air on your hand. This is the basic effect of throttling. In a more extreme case, letting compressed air out of an automobile tire results in air so cold that it often results in ice forming on the nozzle. Thus, forcing the high-pressure liquid refrigerant through the expansion valve in Figure 2.13 results in a liquid on the other side of the valve that is quite cold, even colder than the air inside the refrigerator itself.

This very cold liquid now enters the refrigerator at point D, where it encounters coils referred to as the *evaporator*. In this part of the cycle, from D to A, the cold liquid experiences the comparatively "high" temperature inside the refrigerator. As a result, the refrigerant begins to boil and vaporize, resulting in a significant amount of latent heat of vaporization being transferred from the interior of the refrigerator to the refrigerant at a constant temperature. The resulting vapor exits the refrigerator and completes the cycle at point A.

The refrigeration cycle is presented in a *P–V* diagram in Figure 2.14. From A to B the vapor of the refrigerant is compressed, which decreases the volume and increases the temperature. This process is approximately adiabatic. Next, the refrigerant enters the condenser, where the high-temperature vapor condenses to a liquid, giving off the latent heat of vaporization $|Q_h|$ as the refrigerant moves from B to C. The process of condensation occurs at a constant temperature $T_h$, and hence the process from B to C is isothermal.

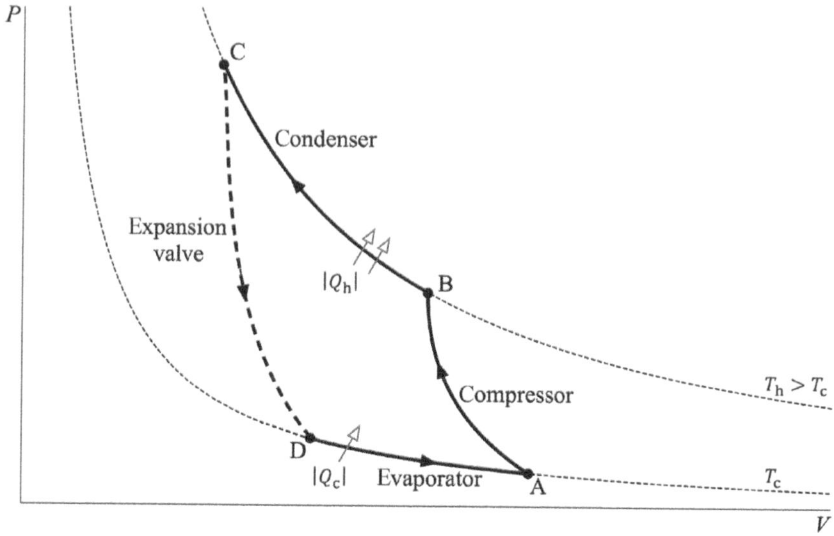


Figure 2.14 The thermodynamic processes of a refrigerator represented in a $P$–$V$ diagram.

Next, the refrigerant encounters the expansion valve. The throttling process that occurs as the liquid passes through the valve is irreversible, and hence it can't be represented by points on a $P$–$V$ diagram. Instead, we indicate the process with a dashed curve that connects the starting point, C, and the ending point, D, at a lower temperature and a slightly greater volume.

The refrigerant is now a low-pressure, low-temperature liquid, and it enters the coils of the evaporator inside the refrigerator. The evaporation process, from D to A, draws the heat energy $|Q_c|$ from inside the refrigerator at the constant temperature $T_c$.

Notice that this $P$–$V$ diagram is the same as that for the heat engine in Figure 2.9, except all the arrows are reversed. This means the cycle is run in the opposite direction, counterclockwise instead of clockwise, indicating that work is done *on* the system, as opposed to work being done *by* the system.

We see that a vapor-compression refrigerator uses the latent heat of vaporization in two different locations, with the net effect of "pumping" heat energy from a cold region to a hot region. Inside the refrigerator the latent heat of vaporization enters the refrigerant, causing it to boil; outside the refrigerator the latent heat of vaporization is given off by the refrigerant as it condenses.

It's worth noting that the phase transitions in a refrigerator have an advantage beyond just moving a lot of heat due to the rather large latent heat of vaporization. Specifically, the temperature in both the condenser and evaporator remain constant because of the phase transitions occurring within them. If the refrigerant in the condenser wasn't undergoing a phase transition, its temperature would fall to a value closer to that of room temperature, and the condenser would be less effective at transferring heat energy. Similarly, without a phase transition the temperature in the evaporator would rise, again reducing the effectiveness of heat energy transfer. Clearly, phase transitions play crucial roles in this important technology.

## 2.6 Problems

2.1 **Heat and temperature.** Suppose 395 J of heat energy is added to 1.5 kg of aluminum and 1.5 kg of copper.
**(a)** Referring to Table 2.1, is the increase in temperature of the aluminum greater than, less than, or equal to the increase in temperature of the copper? Explain your reasoning.
**(b)** Calculate the increase in temperature of the aluminum.
**(c)** Calculate the increase in temperature of the copper.

2.2 **Heat and temperature.** **(a)** How much heat energy must be removed from 2.70 kg of lead to cause a decrease in temperature of 3.45 K?
**(b)** If this same amount of heat is removed from the same mass of ice, is the decrease in temperature greater than, less than, or equal to 3.40 K? Explain your reasoning.
**(c)** Calculate the decrease in temperature of the ice.

2.3 **Heat and temperature.** The temperature of a block of silver increases by 5.75 K when 1250 J of heat energy is added to it. What is the mass of the block?

2.4 **Heat and temperature.** Figure 2.15 shows temperature-versus-heat for four different substances, a, b, c, and d, each with a different starting temperature. The lines, in no particular order, correspond to copper, glass, lead, and ice. Identify the substance corresponding to each letter.

2.5 **Heat and phase transitions.** Find the heat energy required to heat 1.50 kg of ice at $-20.0\,^{\circ}$C to steam at $120\,^{\circ}$C.

2.6 **Heat and phase transitions.** Consider a system consisting of 2.75 kg of ice at an initial temperature of $-20.0\,^{\circ}$C. Find the final state of the system if **(a)** 996,000 J or **(b)** 1,220,000 J of heat is added to the initial state of the system.

2.7 **Heat and phase transitions.** The melting temperature of 24 karat gold is $1063\,^{\circ}$C. Suppose 64,500 J of heat is added to 1.60 kg of gold at $1000\,^{\circ}$C. How much gold melts?

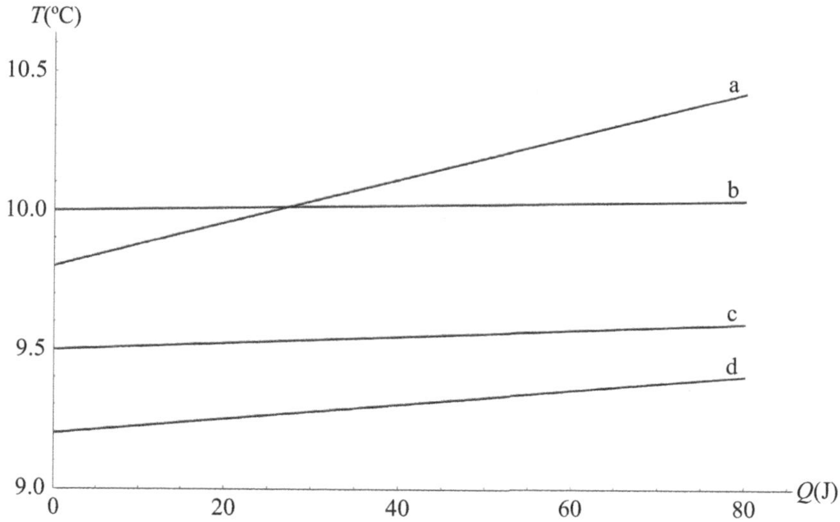


Figure 2.15 Temperature $T$ versus heat $Q$ for four different substances.

2.8 **Clausius–Clapeyron equation.** The latent heat needed to melt ice to water is 335 kJ/kg. The density of ice is 0.915 kg/m$^3$ and the density of water is 1000 kg/m$^3$. Given that the melting temperature of ice is 273.15 K at 1 atm = 0.1013 MPa, what is its melting temperature when the pressure is raised to 110 MPa?

2.9 **Clausius–Clapeyron equation.** The density of solid gold at the melting temperature of 1337 K is 18,300 kg/m$^3$. The latent heat of fusion for gold is 62,800 J/kg, and the density of the resulting liquid is 17,190 kg/m$^3$. Using the linear approximation to the Clausius–Clapeyron equation given in Eq (2.9), find the increase in pressure required to increase the melting temperature of gold by 1.00 K.

2.10 **Clausius–Clapeyron equation.** Solid lead, with a density of 11,340 kg/m$^3$, melts at 600.65 K. The density of liquid lead is 10,678 kg/m$^3$. If it takes 70.65 atm of pressure to increase the melting temperature by 1.00 K, what is the latent heat of fusion of lead?

2.11 **Clausius–Clapeyron equation.** The vapor pressure of benzene at 25.0 °C is 12,700 Pa, and its vapor pressure at 80.0 °C is 86,500 Pa. What is the molar latent heat of vaporization for benzene in kJ per mole?

2.12 **Clausius–Clapeyron equation.** The vapor pressure of carbon tetrachloride at 40.0 °C is 213 torr, and its vapor pressure at 70.0 °C is 612 torr. What is the molar latent heat of vaporization for carbon tetrachloride in kJ per mole?

2.13 **Clausius–Clapeyron equation.** The molar latent heat of vaporization for hexane is 28,900 J/mol. Given that the boiling point for hexane at 1 atm = 760 mmHg is 68.73 °C, what is its vapor pressure at 25.0 °C in mmHg?

2.14 **Clausius–Clapeyron equation.** The boiling temperature of argon at 1.00 atm is 83.8 K. Given that its molar latent heat of vaporization is 1210 J/mol, what is its boiling temperature at 2.25 atm?

2.15 **Clausius–Clapeyron equation.** Chloroform has a vapor pressure of 197 mmHg at 23.0 °C and 632 mmHg at 55.0 °C.

**(a)** What is the molar latent heat of vaporization for chloroform?

**(b)** What is the boiling temperature of chloroform at atmospheric pressure (1 atm = 760 mmHg)?

2.16 **Heat engines.** Figure 2.16 shows two different heat engines, **(a)** and **(b)**, taking in the same amount of heat, $|Q_h|$, but doing different amounts of work, $|W|$, and giving off different amounts of exhaust heat, $|Q_c|$. Is the efficiency of heat engine **(a)** greater than, less than, or equal to the efficiency of heat engine **(b)**? Explain your reasoning.

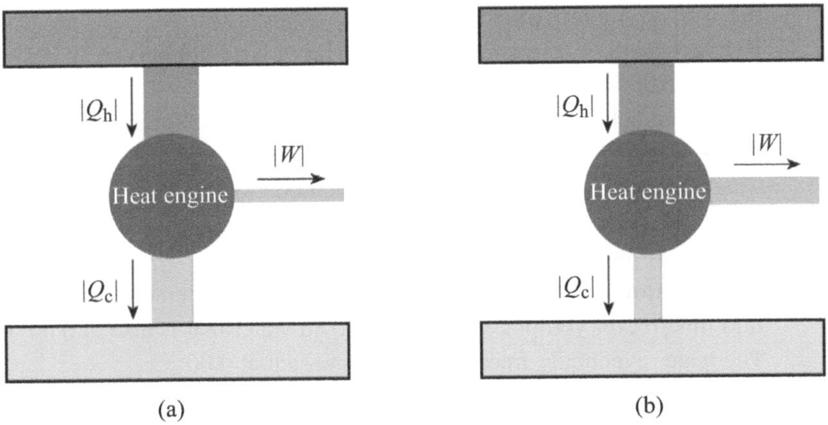


Figure 2.16 Two heat engines of different efficiencies.

2.17 **Heat engines.** Heat engines 1 and 2 do the same amount of work. Engine 1 takes in 100 J of heat, and engine 2 takes in 125 J of heat. Is the efficiency of heat engine 1 greater than, less than, or equal to the efficiency of heat engine 2? Explain your reasoning.

2.18 **Heat engines.** A heat engine absorbs 6550 J of heat at high temperature and performs 1240 J of work.

**(a)** What is the efficiency of this engine?

**(b)** How much heat does the engine give off at low temperature?

**(c)** Using the Kelvin temperature scale, what is the ratio of the low temperature to the high temperature, $T_c/T_h$?

2.19 **Heat engines.** A heat engine absorbs 1.50 times as much heat as it exhausts, and the work it does per cycle is 120 J.
**(a)** What is the efficiency of this engine?
**(b)** How much heat does the engine absorb per cycle?
**(c)** How much heat does the engine exhaust per cycle?

2.20 **Heat engines.** Consider a heat engine running on a Carnot cycle with an efficiency of 0.275. The high temperature for this engine is 512 K.
**(a)** What is the low temperature of this engine?
**(b)** To increase the efficiency of this engine, should the low temperature be increased or decreased? Explain why.
**(c)** Find the low temperature of this engine that gives an increased efficiency of 0.390.

2.21 **Refrigerators.** Over a given period of time, the refrigerator in your kitchen does 325 J of work and removes 78 J of heat from its interior.
**(a)** How much heat does the refrigerator exhaust into the kitchen during this time?
**(b)** What is the refrigerator's coefficient of performance?

2.22 **Refrigerators.** A refrigerator cools 2.24 kg of water initially at 20.0 °C to ice at −5.00 °C.
**(a)** How much heat must be removed from the water for this to occur?
**(b)** How much electrical energy must the refrigerator use to freeze the water if its coefficient of performance is 4.00?
**(c)** How much heat is exhausted by the refrigerator into the kitchen during this process?

2.23 **Refrigerators.** A freezer has a constant interior temperature of 0.00 °C. Into this freezer you place 0.454 kg of cold water that is also at 0.00 °C. The water eventually freezes and becomes ice at 0.00 °C.
**(a)** How much heat must be removed from the water to cause it to freeze?
**(b)** What is the change in entropy of the water as it freezes?
**(c)** If the coefficient of performance of the freezer is 4.70, how much heat does the freezer exhaust into the kitchen?
**(d)** What is the increase in entropy of the kitchen, given that it is at the constant temperature of 22.0 °C.
**(e)** What is the change in entropy of the universe as a result of this process?

2.24 **Refrigerators.** The food compartment of a refrigerator is kept at 5.00 °C. To do so, the refrigerator must remove heat from the compartment at the rate of 420 kJ per minute. If the refrigerator needs 2.50 kW of power to keep the food compartment cool, what is the refrigerator's coefficient of performance?

# 3

# The Phases and Phase Transitions of Water

Water is an amazing substance. Not only does it cover 71% of the earth, where it can be found in each of its three common phases, but all living creatures contain water within their bodies. Humans, in particular, are approximately 60% water. In this chapter, we explore some of the characteristics of water that are most important to life on Earth.

## 3.1 The Phases of Water

We start with the basic unit of water – the water molecule. This molecule is composed of one oxygen atom and two hydrogen atoms, as shown in Figure 3.1. This results in the well-known chemical formula $H_2O$. The molecular weight of a water molecule is equal to the molecular weight of oxygen (16 g/mol) plus two times the molecular weight of hydrogen (1 g/mol), giving a total of 18 g/mol.

Compared to air – which consists mostly of nitrogen molecules, $N_2$ (28 g/mol), and oxygen molecules, $O_2$ (32 g/mol) – water molecules are considerably lighter. Thus, on a humid day, with significant amounts of water vapor mixed in with the nitrogen and oxygen molecules, air is less dense than on a dry day. This may seem counterintuitive because humid air can often feel "heavy" and oppressive. We discuss issues related to humidity in the atmosphere near the end of Section 3.2.

Water is a polar molecule, meaning that one end is negatively charged, as indicated with minus signs (– –) in Figure 3.1, while the other end is positively charged, indicated with the positive signs (+, +). Note that the oxygen atom is negatively charged and the hydrogen atoms are positively charged. The molecule itself is electrically neutral. The polar nature of the molecule makes it particularly effective as a solvent because the water molecules can attach

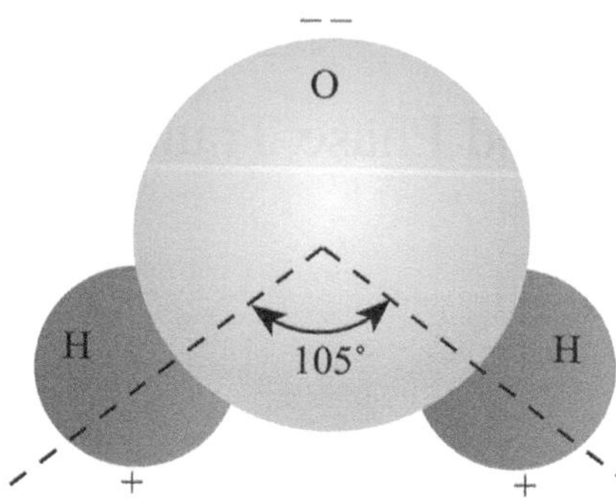


Figure 3.1 Schematic representation of a water molecule, $H_2O$.

themselves to other molecules by electrostatic attraction to facilitate bringing them into solution.

Note also that the water molecule has a "bent" shape, with an angle of about 105° between the hydrogen atoms. This becomes important when the molecules bond to one another to form a solid phase, as we discuss later in this section.

Individual water molecules have their own particular properties, but when large numbers of them are working together as a group, they can take on new, *emergent* properties not associated with the molecules themselves. These new properties can be described in terms of different phases of matter. We address the properties of the three most common phases of water next.

## Water Vapor: Steam

Water exists as a vapor, or gas, at high temperatures and low pressures. We refer to this phase as *steam*. The properties of steam are really quite simple because the molecules in this phase move about freely, with essentially no contact or interactions with one another, as indicated in Figure 3.2.

Specifically, if you have a container with $n$ moles of steam at temperature $T$, the molecules will occupy the entire volume $V$ of the container. As the molecules bounce off the walls they produce an outward pressure $P$. Because of the lack of intermolecular interactions, the behavior of the steam is well described by the ideal gas law:

$$PV = nRT.$$

In this expression, $R$ is the universal gas constant: $R = 8.3145\,\text{J/(mol}\cdot\text{K)}$.

Steam is a low-density phase, with a density of about $0.590\,\text{kg/m}^3$, and is invisible – just like air. We often think of steam as being white and puffy, like the exhaust from a steam locomotive, but what we're actually seeing in a case

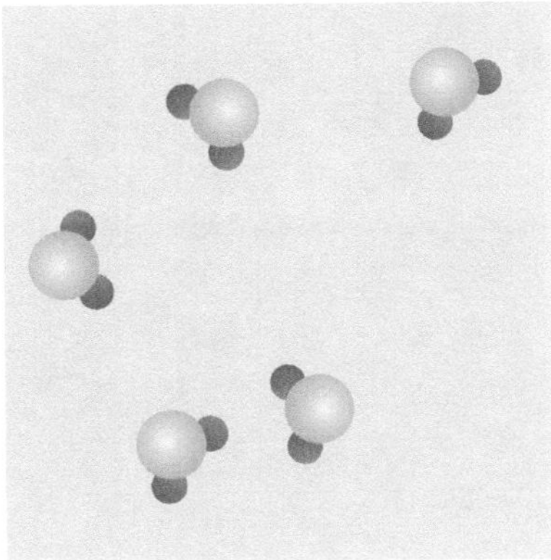

Figure 3.2 Water vapor (steam) consists of widely separated molecules with low density.

like this is steam vapor condensing into small liquid droplets, which *are* visible. We turn now to the liquid phase of water.

## Liquid Water

Liquid water – or simply *water* in common usage – is a phase that exists over a relatively small range of temperatures and pressures between the vapor and solid phases. In the liquid phase of water, the molecules are relatively closely packed, as indicated in Figure 3.3. The density of this phase is fairly high, at about 1000 kg/m$^3$, which is roughly 1600 times the density of steam.

The fact that the molecules in liquid water are quite close to one another means that their interactions are significant. In fact, in liquid water the molecules are constantly forming and breaking weak hydrogen bonds with one another. If you put a quantity of liquid water in a container, it is filled from the bottom up. The liquid is splashy, sloshes around in the container, and is virtually incompressible.

Raising the temperature of liquid water results in it boiling to produce steam. On the other hand, if liquid water is cooled, it freezes to form solid crystals of ice. We consider ice next.

## Solid Water: Ice

Solid water – more commonly known as *ice*, of course – is the phase that exists at low temperatures and high pressures. In this phase, water molecules form strong, permanent hydrogen bonds with one another. Because of the specific angle between the hydrogen atoms in an $H_2O$ molecule, as indicated in

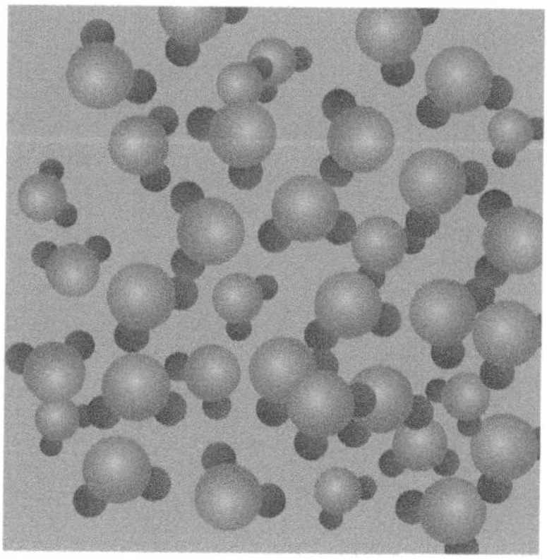

Figure 3.3 Water molecules in liquid water. Note the relatively high density of the molecules in this phase.

Figure 3.1, the bonded molecules form a hexagonal crystalline structure, as shown in Figure 3.4.

Now, in most substances lowering the temperature, or raising the pressure, results in the substance becoming increasingly dense. It makes sense, for example, that cooling a substance increases its density as the molecules move about more slowly and pack more closely together. What substance could possibly behave differently than this? Well, that would be water, of course. It is one of only a small number of materials known to expand as it cools near the freezing point, and thus to have a solid phase that is less dense than the corresponding liquid phase.

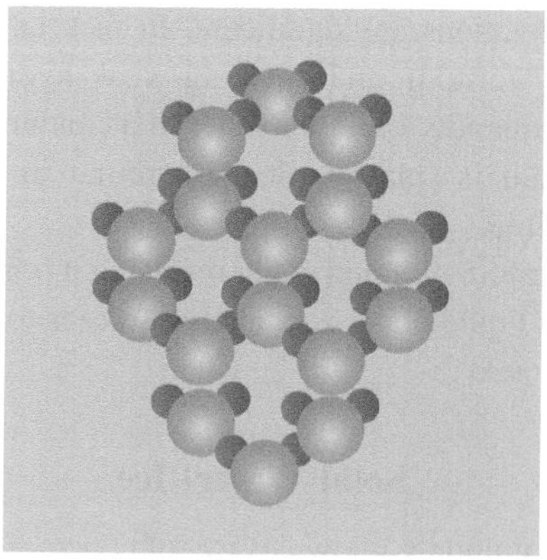

Figure 3.4 Water molecules in solid water (ice). Notice the open structure of the hexagonal ice crystal, resulting in a relatively low density compared with the liquid phase in Figure 3.3.

So, why does water expand when cooled? We can see how this comes about in Figure 3.4. As the water molecules link together to form a hexagonal structure, they leave *open spaces* in the centers of the hexagons. This lowers the overall density. Thus, the density of liquid water initially increases with cooling from room temperature, as one would expect, but at about 4 °C the molecules begin to form hexagonal structures. As a result, the density starts to decrease. By the time the water freezes solid at 0 °C the density has decreased to about 917 kg/m$^3$, about 10% less than the density of liquid water. This is why ice floats in water. It's worth noting that what is incredible is not that 90% of an iceberg is below the water, as is often stated, but that 10% – or any amount at all – is above the water's surface.

At still lower temperatures and higher pressures, other crystal forms of ice can be produced in the lab, each with a slightly different structure. The most common type of ice, the one we encounter in everyday life, is shown in Figure 3.4. This structure is referred to as "ice $I_h$" (pronounced "ice one h") indicating that it was the first form of ice to be described (I), and that its structure is hexagonal (h).

A purely fictitious form of ice, known as "ice-nine," appears in Kurt Vonnegut's science fiction novel *Cat's Cradle*. This type of ice is imagined to be super stable, and to act as a seed crystal that causes liquid water to suddenly freeze to produce more ice-nine. As a result, a person who ingests ice-nine is killed instantly. In the climax of the novel Earth's oceans are accidently frozen solid by ice-nine, resulting in a doomsday scenario for the planet. Fortunately, there is no such thing as ice-nine in the real world.

The six-fold symmetry of water molecules linking together in ice crystals manifests itself macroscopically in the form of six-sided snowflakes. Examples are shown in Figure 3.5. No two snowflakes are identical, of course, because each one grows at its own pace as it falls toward the ground, passing through regions of different temperature and humidity. They all share a six-fold symmetry, however, showing an infinite variety based on a shared theme.

Now that we've considered the three common phases of water, we turn to the phase diagrams that show how these phases relate to one another in terms of temperature and pressure.

## 3.2 The Phase Diagram of Water

We show a detailed *P–T* phase diagram for water in Figure 3.6. Included in the diagram are some special points and dashed lines. For example, the middle dashed horizontal line corresponds to atmospheric pressure, as experienced in our everyday lives. On this line we see that solid ice melts to liquid water at a

(a)

(b)

Figure 3.5 Six-sided snowflakes formed from solid water. (a) An actual snowflake. (b) A diagrammatic representations of typical snowflake shapes. Note that they all have a six-fold symmetry. Source: (a) Detached Retina Imaging/Moment/Getty Images; (b) Hafid Firman Syarif/iStock/Getty Images Plus.

temperature of precisely 0 °C, just below 0.01 °C, as given by the definition of the Celsius scale. Liquid water persists as the temperature is increased until the boiling point is encountered at 100 °C – again, by definition. At this point, liquid water boils to produce gaseous steam.

The lower dashed horizontal line in Figure 3.6 corresponds to the *triple point* of water. This point occurs at a pressure of 0.006 atm and a temperature of 0.01 °C, and is the *one point* in the phase diagram where all three phases exist at the same time. At the triple point, which is readily prepared in the lab, we can have a system in equilibrium with ice cubes floating in boiling water producing steam.

The top dashed horizontal line in Figure 3.6 corresponds to the *critical point* of water, which occurs at a pressure of 218 atm and a temperature of 374 °C. Beyond this point there is no longer a distinction between the liquid and vapor phases, as discussed in Chapter 1. We'll consider this point in more detail later in this section.

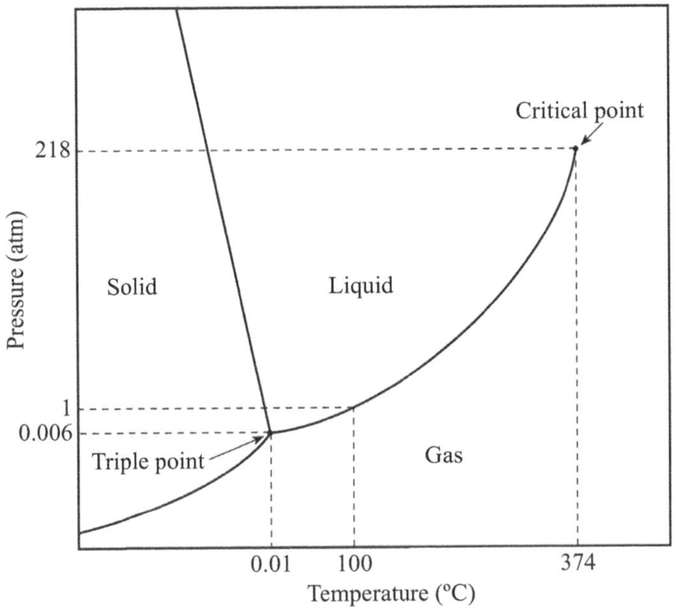


Figure 3.6 The *P–T* (pressure–temperature) phase diagram for water. Notice that the pressure is measured in atm, and the temperature is measured in degrees Celsius.

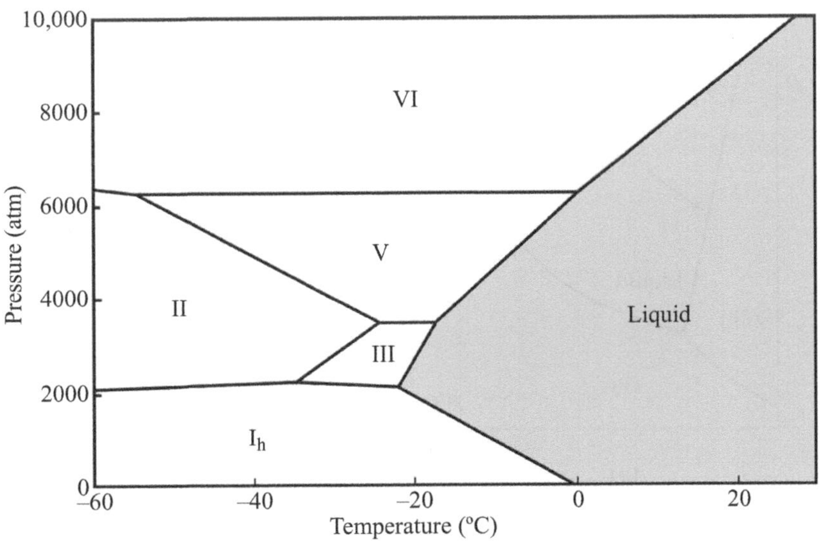


Figure 3.7 Phase diagram for a variety of ice phases.

An enlargement of water's phase diagram showing a number of different ice phases is presented in Figure 3.7. Notice that these phases occur at low temperature and extremely high pressures – on the order of several thousand atmospheres.

Figure 3.7 shows just 5 different ice phases, but there are at least 19 phases of ice known to exist, each with its own slightly different crystal structure. The ice phases of water continue to be a research topic.

## The Solid–Liquid Phase Boundary

The most striking feature of the phase diagram of water, as shown schematically in Figure 3.8(a), is the "backward" or *negative slope* of the phase boundary between the solid and liquid phases. This is in contrast to the *positive slope* for the solid–liquid phase boundary in most cases, an example of which is shown in Figure 3.8(b) for carbon dioxide, $CO_2$. Notice that the slopes of all of the *P–T* phase boundaries are positive in typical phase diagrams.

Recall from Section 2.4 that the rate of change of pressure with temperature for a solid–liquid phase boundary is given by the following form of the Clausius–Clapeyron equation (Eq. (2.9)):

$$\Delta P = \left[\frac{l_f \rho_{\text{solid}} \rho_{\text{liquid}}}{T\left(\rho_{\text{solid}} - \rho_{\text{liquid}}\right)}\right] \Delta T.$$

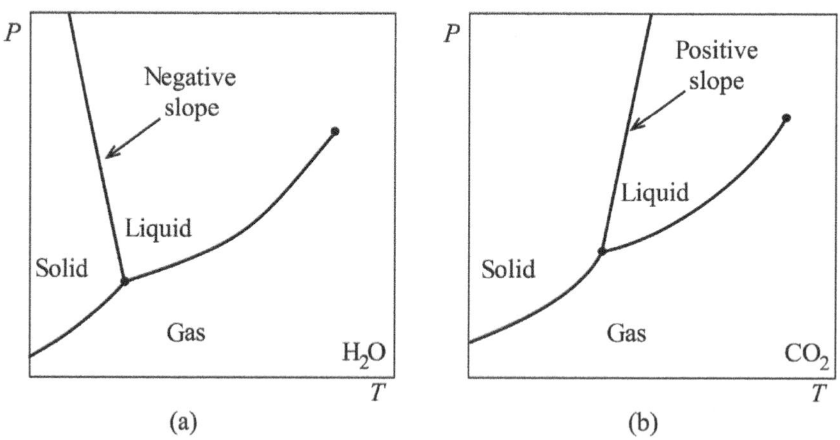


Figure 3.8 A comparison of the *P–T* phase diagram of (a) water and (b) carbon dioxide. Notice that the slope of the solid–liquid phase boundary is negative in the case of water.

We see that in the case of ice and water, where the density of the solid (ice) is less than the density of the liquid (liquid water), the denominator of this equation is negative, giving a negative slope to the phase boundary. It's interesting to see how the overall shape of a phase diagram results directly from the density of its individual phases.

Now that we know that the slope of the phase boundary is negative, let's calculate its numerical value. For water, we have $\rho_{\text{solid}} = 917\ \text{kg/m}^3$, $\rho_{\text{liquid}} = 1000\ \text{kg/m}^3$, and $l_f = 335{,}000\ \text{J/kg}$. The temperature of the phase transition is 0 °C, or $T = 273.15$ K. With these values we find the following relationship between $\Delta P$ and $\Delta T$:

$$\Delta P = -(13.5\ \text{MPa/K})\Delta T.$$

Note that atmospheric pressure (1 atm) is given by 1 atm = 0.101 MPa. Thus, it follows that

$$\Delta P = -\left(134\ \frac{\text{atm}}{\text{K}}\right)\Delta T.$$

So, it takes an increase in pressure of 134 atm to decrease the melting temperature of ice by one Kelvin.

A controversial application of this relation is to ice skating, where it is often said that the pressure produced by the blade of an ice skate exerts enough pressure to melt the ice below it, which in turn facilitates an effortless gliding motion. The numbers make this doubtful, however, and the more accepted explanation is that friction between the skate's blade and the ice is responsible for the melting. In any case, ice is slippery, while other solids are not, and increased pressure does cause ice to melt.

Another perspective on this situation is provided by the sport of curling. Here the curling "stone" is large compared to the blade of an ice skate, and has a mass much less than a person. Thus, the pressure exerted by a curling stone is much less than the pressure exerted by the blade of an ice skate. Still, the stone glides over the surface of the ice with a nearly frictionless motion. In fact, the competitors control the distance the stone slides by scrubbing the ice in front of it to melt the surface, thereby either increasing or decreasing the distance it covers before stopping. Clearly friction, rather than pressure-induced melting, is the primary factor in this case.

The fact that ice has a smaller density than liquid water means that ice expands in volume when it freezes. This is why water pipes burst in the winter in northern climes, when the liquid water inside the pipes freezes, expands, and applies such a great outward pressure that it can burst open a metal pipe. Similarly, liquid water that seeps into cracks in rock formations expands when it freezes in cold weather, and applies an enormous force to the rock, causing it to split and break. Thus the expansion of ice is an important factor in erosion, especially in climates with significant freezing in winter.

The expansion of ice is also responsible for the phenomenon of frost heaving, in which freezing causes rocks in a field to be slowly elevated with each freeze-and-thaw cycle, thus slowly moving rocks upward to cover a field previously clear of rocks.

Finally, the lower density of ice compared to that of liquid water means that lakes in northern climates freeze from the top down. When the weather turns cold, a layer of ice will form on the lake's surface. The ice coating partly insulates the water below, helping to keep the rest of the lake from freezing. The densest water, at a temperature of 4 °C, as mentioned previously, sinks to the bottom of the lake. As a result, organisms that live in the lake can survive in relatively warm liquid water deep in the lake throughout the winter. This would be impossible if the lake froze from the bottom up, resulting in a completely solid mass of ice, which might not even melt fully during the following summer.

Clearly, the lower density of ice compared to that of liquid water has many significant implications for the natural world in addition to determining the slope of the solid–liquid phase boundary in water's $P$–$T$ phase diagram.

## The Liquid–Gas Phase Boundary

We now turn to the liquid–gas version of the Clausius–Clapeyron equation, as derived in Section 2.4. In particular, recall Eq. (2.14), which gives the $P$-versus-$T$ liquid–gas phase boundary for water:

$$P = (1.00\ \text{atm})e^{-(4897\ \text{K})\left(\frac{1}{T} - \frac{1}{373.15\ \text{K}}\right)}.$$

This equation is written in terms of the absolute temperature $T$. For convenience, when plotting in terms of the Celsius temperature, $T_C$, we use the following equivalent relation:

$$P = (1.00\ \text{atm})e^{-(4897\ \text{K})\left(\frac{1}{273.15 + T_C} - \frac{1}{373.15\ \text{K}}\right)}. \tag{3.1}$$

This is a rather unusual function of temperature. It starts off with an upward curvature (positive second derivative), followed at higher temperatures by a downward curvature (negative second derivative), and finally an asymptotic approach to a large but finite value. The region of physical interest for the liquid–gas phase boundary in water is the low-temperature region with an upward curvature.

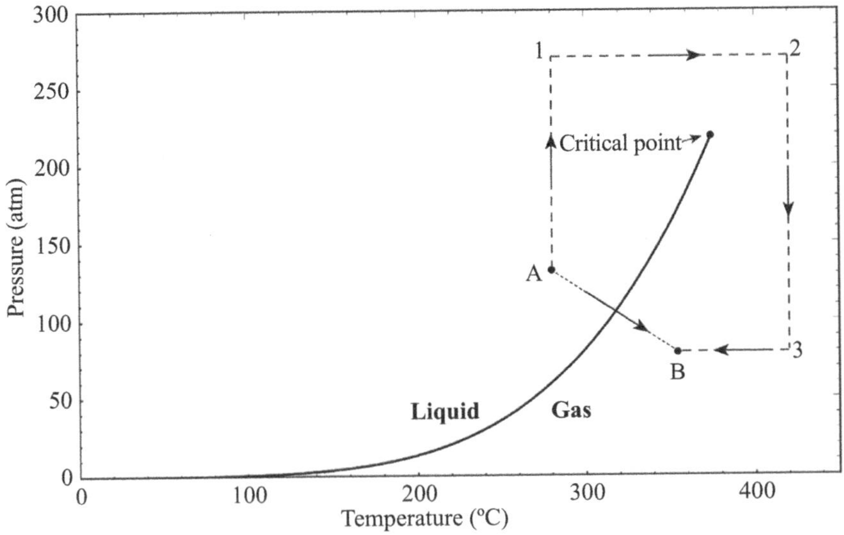


Figure 3.9 The liquid (A) and gas (B) phases of water can be connected by a phase transition by going directly from A to B, or without a phase transition by going around the critical point, through points 1, 2, and 3.

For example, Figure 3.9 shows a plot of Eq. (3.1) for the range of temperatures from 0 °C to 450 °C. This range extends from the triple point of water at 0.01 °C and 0.006 atm, past the critical point at 374 °C and 218 atm. Point A in this plot is a splashy liquid, and point B is a gas. Changing the pressure and temperature along the straight line from point A to point B results in a phase transition when the path crosses the phase boundary, with a vigorous boiling of the liquid water and the production of vapor.

On the other hand, consider connecting A and B by the path that goes through point 1 (at 280 °C and 270 atm), point 2 (at 420 °C and 270 atm), and point 3 (at 420 °C and 80 atm). The same initial and final states are connected by this path, but no phase transition occurs. In effect, the liquid and gas phases can be considered as different aspects of a single phase, referred to as the fluid phase, as mentioned previously.

In Figure 3.10 we zoom in on the range of temperatures from 30 °C to 140 °C. As in Figure 3.6, the middle dashed lines show the boiling temperature of water (100 °C) at atmospheric pressure (1 atm). In addition, we note that boiling occurs at a higher or lower temperature as the pressure is increased or decreased, respectively.

As a specific example, consider using boiling water to cook a meal at high altitude. In Figure 3.10 we show the reduced pressure found at 10,000 ft of

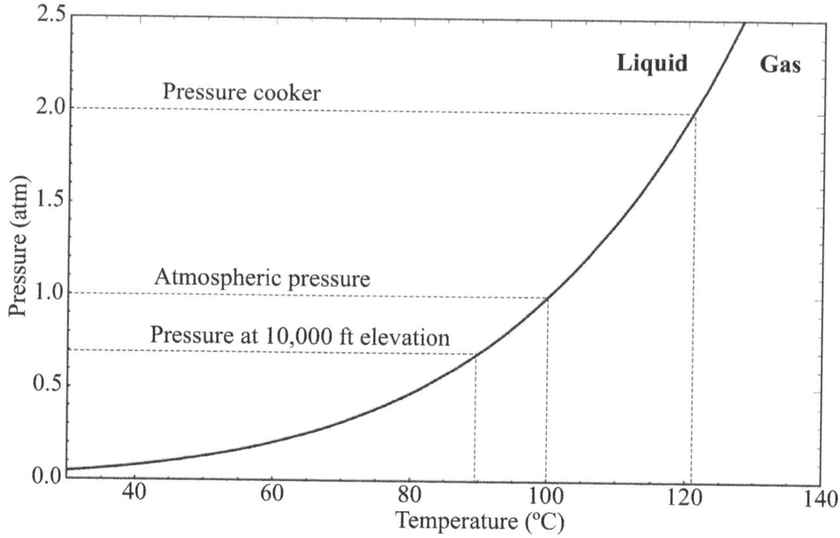


Figure 3.10 Phase boundary between liquid and gaseous water. When the pressure is 2 atm, as in a pressure cooker, the boiling temperature is slightly more than 120 °C. When the pressure is less than 1 atm the boiling temperature is less than 100 °C.

elevation, about 70% of that at sea level, and the corresponding boiling temperature of water. What the Clausius–Clapeyron equation tells us is that water boils at that altitude at a tepid 90 °C, which really isn't hot enough to cause raw food to cook properly.

Charles Darwin's book *Voyage of the Beagle*, which recounts his trip around the world, describes his experiences exploring the Andes mountains with local guides. He tells of trying to cook potatoes while camping on March 21, 1835, at an altitude of roughly 10,000 ft. He wrote the following:

> Our potatoes, after remaining for considerable time in the boiling water, were nearly as hard as ever. The pot was left on the fire all night and the next morning it was boiled again, but yet the potatoes were not cooked. My two companions discussed the cause and came to the simple conclusion that the cursed pot (which was a new one) did not choose to boil the potatoes.

That's one explanation. Darwin's conclusion was different, however, and correct:

> At the place where we slept water necessarily boiled, from the diminished pressure of the atmosphere, at a lower temperature than it does in a less lofty country.

This was a novel observation in the early nineteenth century, but is much more commonly known today.

On the other hand, increasing pressure causes the boiling temperature to rise, and this results in a more rapid cooking process. An example is a pressure cooker, which is a sealed vessel that typically operates at a pressure of about 2 atm. As shown in Figure 3.10, this results in cooking at roughly 120 °C, which significantly shortens cooking times.

A geothermal geyser is an example of a "natural" pressure cooker. Groundwater seeps into underground chambers where the pressure of the water column, enhanced by the constrictive channels leading to the surface, can result in pressures of 9 or 10 atm. At these pressures, water can become superheated and remain liquid to temperatures in the neighborhood of 400 °F. When some of the water near the surface reaches its local boiling point it can flash into steam, which reduces the pressure on the water deeper down. As the pressure drops, the superheated water suddenly finds itself above its new boiling point, and it flash-boils to steam. Since steam takes up about 1600 times more volume than liquid water, the result is an explosive eruption of steam and water out of the vent of the geyser.

## Relative Humidity and Dew Points

Zooming in on the phase diagram for water once again, to a temperature range of 0 °C to 40 °C, we find the phase diagram shown in Figure 3.11. This is the

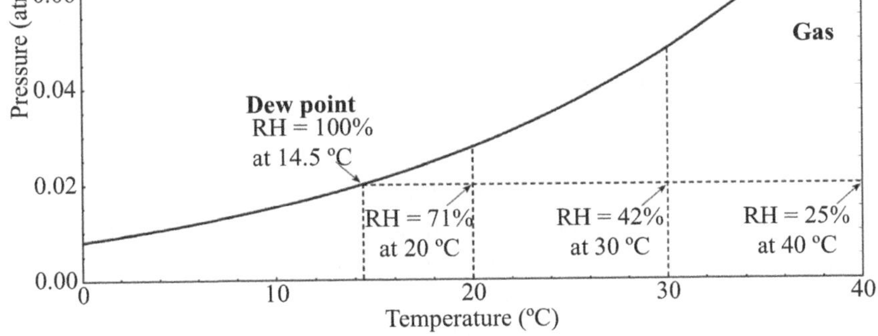


Figure 3.11 Comparing relative humidity, RH, and the dew point.

range of temperatures typically encountered in everyday life. The pressures in this temperature range are on the order of a tenth of an atmosphere and less.

Now, there is always some water vapor in the atmosphere, which we refer to as *humidity*, and the atmospheric pressure we experience is, by Dalton's law, the sum of the partial pressures due to air and to water. When the partial pressure of water vapor is below the curve in Figure 3.11, the water remains in the vapor, or gaseous, phase in the atmosphere. On the other hand, suppose the partial pressure of water is increased at constant temperature until it hits the curve. At this point we say that the water vapor has reached its *saturation point*, or more precisely, its *saturation pressure*, and it begins to condense to form liquid droplets of dew.

With this in mind, we define the *relative humidity*, RH, of air as follows:

$$\mathrm{RH} = \frac{\text{partial pressure of water vapor}}{\text{saturation pressure of water vapor}} \times 100\%.$$

For example, consider the temperature of 40 °C in Figure 3.11. At this temperature the saturation pressure of water vapor is about 0.08 atm. It follows that one-quarter of this value, which is roughly 0.02 atm, corresponds to a relative humidity of 25 % – that is, RH = 25% – as indicated in the plot.

If the temperature is now reduced, with the water partial pressure remaining constant, we find that at 30 °C the saturation pressure of water has dropped to 0.048 atm, and hence the relative humidity is now RH = (0.02/0.048) × 100% = 42%. Similarly, at 20 °C the saturation pressure is 0.028 atm, and the relative humidity is RH = 71%. Finally, at 14.5 °C we hit the saturation curve and the relative humidity has risen to 100%. This is referred to as the *dew point* because at this temperature and pressure liquid droplets of dew begin to condense on surfaces exposed to the air.

The dew point is particularly important because it determines how we experience the humidity in the air. This is indicated on the chart showing the *dew-point temperature range* in Figure 3.12. For example, if the dew point is below 10 °C (50 °F), the air feels dry to us, and our skin can become brittle and crack. We experience dew points at temperatures between 10 °C and 15 °C (50 °F to 59 °F) as completely comfortable. This is basically the ideal range. Dew points between 15 °C and 20 °C (59 °F to 68 °F) are experienced as "muggy"; that is, noticeably humid. Finally, dew points in excess of 20 °C (68 °F) are felt as truly "oppressive," meaning that when you go out on such a day you feel as if the air is heavy and pressing down on you. In fact, as mentioned earlier, the air is actually less dense when the humidity is higher, but here we are concerned with the physiological feeling we experience. When the dew point is that high, it's hard for our bodies to shed heat into the

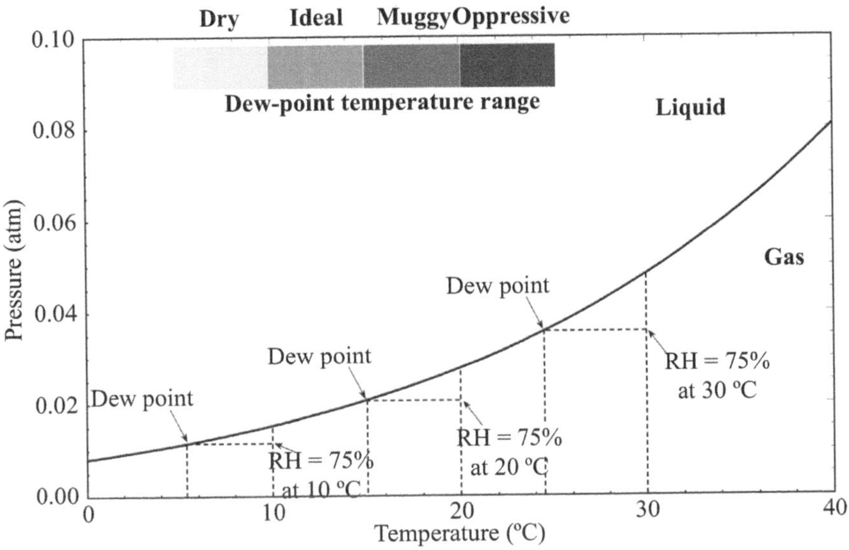


Figure 3.12 Dew points determined by temperature and relative humidity. The dew-point temperature range indicates the experience associated with various dew-point temperatures.

atmosphere through perspiration, and we can succumb to overheating and heat exhaustion.

Also shown in Figure 3.12 is the connection between relative humidity and dew points. For each of the temperatures 30 °C, 20 °C, and 10 °C, we show the partial pressure of water vapor corresponding to a relative humidity of 75%. For example, a relative humidity of 75% at 30 °C corresponds to a dew point of just under 25 °C. This is well into the oppressive range of humidity. On the other hand, the same relative humidity of 75% at 20 °C corresponds to a dew point of about 15 °C, which is at the upper limit of the ideal range of humidity. Finally, 75% relative humidity at 10 °C yields a dew point of about 5.5 °C, which is well into the dry range of humidity. Thus, the same relative humidity can produce a wide range of different experiences depending on the corresponding dew point.

## 3.3 Hurricanes as Heat Engines

At its core, a hurricane is a gigantic, atmospheric heat engine. The high-temperature reservoir of the engine is the warm water of the ocean, and the low-temperature reservoir is the cool air at high altitude. A schematic

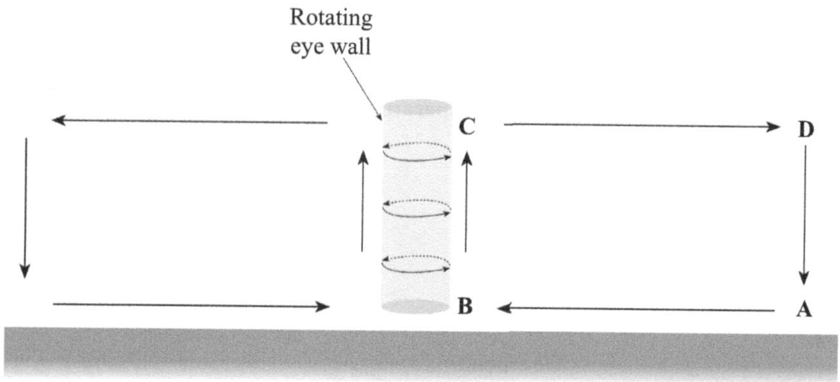


Figure 3.13 Basic elements of a hurricane.

illustration of the way air circulates through a hurricane is shown in Figure 3.13. The rotation of the earth results in the counterclockwise rotational motion of a hurricane in the northern hemisphere, but here we focus on the thermodynamics of the hurricane rather than its rotational motion.

We start at point A in Figure 3.13, where a packet of relatively cool air near the surface of the ocean moves toward the eye, where the air pressure is low. As it moves along, the packet absorbs heat from the warm waters of the ocean. By the time the packet reaches point B it is considerably warmer. In fact, it is actually warmer than the surrounding air due to its close proximity to the warm ocean waters, and so it begins to rise, much like a hot-air balloon.

We note that the packet of air not only warms up as it goes from A to B, but it also picks up a lot of water vapor, so it is much more humid at B than at A. As it rises from B to C, it begins to cool with altitude, but – and here's the key part – it *cools less than the surrounding air* because the water vapor it contains begins to condense and release the latent heat of vaporization. Thus, the packet stays warmer than its surroundings due to the phase transition of the water vapor, and it continues to rise higher than it would otherwise.

At the top of the hurricane, at point C, the air packet starts to release heat to the cool air at high altitude. The air then moves out to point D, by which point it has cooled to the high-altitude temperature. As it drops from D to A the air warms up again, so that when it arrives at point A its temperature is moderate, and is ready to increase once again as it flows back to point B.

We represent this in terms of a heat engine in Figure 3.14. From point A to point B the system is following an approximately isothermal process at the

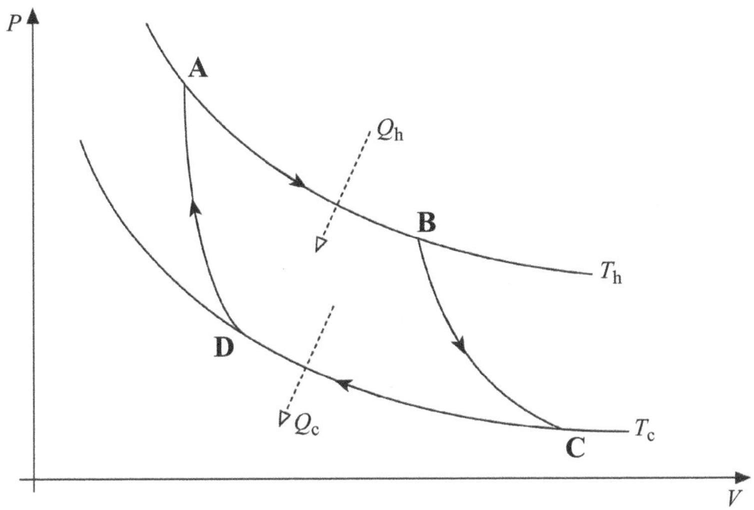


Figure 3.14 A schematic heat engine corresponding to the basic operation of a hurricane.

warm temperature of the ocean. As it does so, the air pressure decreases, and the volume of the air packet increases. During this part of the cycle the air packet absorbs the heat energy $Q_h$ from the ocean. Because the specific heat capacity of water is so abnormally large compared to other substances, it follows that an increase in water temperature by even a degree Celsius results in a great increase of heat energy available to heat the air packet.

From point B to point C the air packet follows an approximately adiabatic process, and its temperature decreases. The temperature decreases less than would otherwise be the case because of the heat released by the latent heat of the water vapor. Because the temperature of the heat packet decreases more slowly than the air around it, whose temperature drops at a rate referred to as the *lapse rate*, the packet continues to rise to great altitude.

On the process from point C to point D, which is approximately isothermal, the system releases the heat $Q_c$ to the upper atmosphere. Finally, as the air packet drops back to the ocean its temperature increases, as indicated by the process connecting points D and A.

The net effect is that the hurricane is basically a heat engine operating between high and low temperatures. The mechanical work done by the engine appears in the form of the kinetic energy of the air, resulting in very high-speed winds. The latent heat due to the phase transition from vapor to liquid provides much of the energy that fuels this process.

## 3.4 Problems

3.1 **Interpreting phase diagrams.** Referring to Figure 3.6:
**(a)** What is the phase of water at 200 atm and 200 °C?
**(b)** What is the phase at 200 atm and 400 °C?

3.2 **Interpreting phase diagrams.** Referring to Figure 3.6:
**(a)** What is the sequence of phases at 150 atm as the temperature is increased from –50 °C to 500 °C?
**(b)** What is the sequence of phases at 0.10 atm as the temperature is increased from –50 °C to 500 °C?

3.3 **Interpreting phase diagrams.** Refer to Figure 3.9 for this problem. Does the density of water increase, decrease, or stay the same from **(a)** point A to point 1, **(b)** point 1 to point 2, **(c)** point 2 to point 3, and **(d)** point 3 to point B? Explain your reasoning.
**(e)** Does the density of water increase, decrease, or stay the same as the system goes from point A to point B? Explain your reasoning.

3.4 **Clausius–Clapeyron equation.** Using Eq. (3.1), find the pressure of the liquid–gas phase transition at 300 °C.

3.5 **Clausius–Clapeyron equation.** Using Eq. (3.1), find the temperature of the liquid–gas phase transition at 151 atm.

3.6 **Dew point.** On a given day the temperature is 30 °C and the dew point is 20 °C.
**(a)** What is the relative humidity on this day?
**(b)** What is the experience of the humidity on this day? (Refer to Eq. (3.1) and Figure 3.12.)

3.7 **Dew point.** On a given day the temperature is 35 °C and the relative humidity is 24%.
**(a)** What is the dew point on this day?
**(b)** What is the experience of the humidity on this day? (Refer to Eq. (3.1) and Figure 3.12.)

3.8 **Dew point.** On a given day the relative humidity is 26% and the dew point is 12 °C.
**(a)** What is the temperature on this day?
**(b)** What is the experience of the humidity on this day? (Refer to Eq. (3.1) and Figure 3.12.)

# 4

# The van der Waals Equation of State

We've learned a lot about phase transitions to this point, including aspects related to specific heats and latent heats, as well as equations describing the shape of phase boundaries. We turn now to the first mathematical description for the actual mechanism behind a liquid–gas phase transition. This is due to J. D. van der Waals, who introduced an equation of state in his 1873 PhD dissertation that replaces the familiar ideal-gas relation. The simple changes made by van der Waals, combined with a key insight provided by J. C. Maxwell in 1875, enables his equation of state to describe the behavior of real systems as they undergo a phase transition. In 1910, van der Waals received the Nobel Prize in physics for his contribution to the thermodynamics of liquids and gases. We devote this chapter to a detailed study of the van der Waals equation of state.

## 4.1 Exploring the van der Waals Isotherm

We begin with the ideal gas, which was the starting point for van der Waals' contributions. Recall that the behavior of an ideal gas – consisting of $n$ moles of noninteracting point molecules – is described by the following equation of state relating pressure, $P$, volume, $V$, and temperature, $T$:

$$PV = nRT.$$

In this expression, $R = 8.314\,\mathrm{J/(mol \cdot K)}$ is the universal gas constant. Rearranging slightly, we find that

$$P = \frac{nRT}{V} = \frac{RT}{v}.$$

Here we have introduced $v = V/n$ as the molar volume.

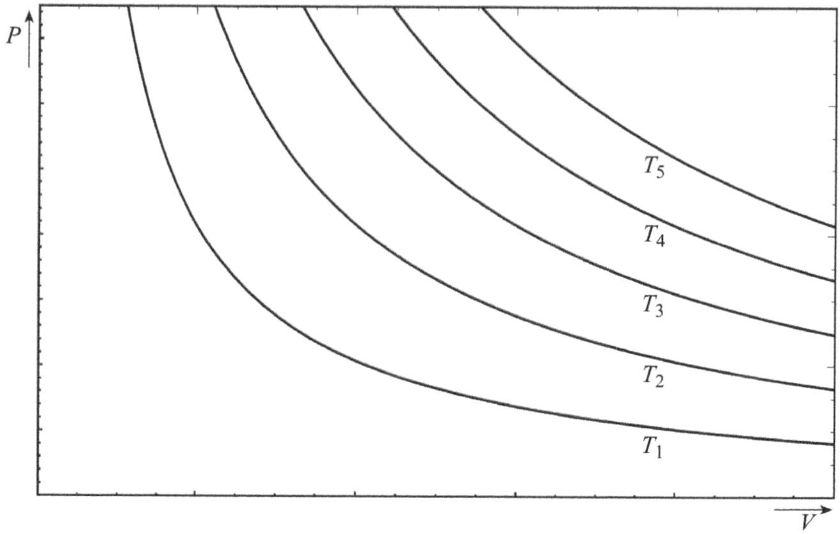


Figure 4.1 Ideal-gas isotherms. Each curve is a hyperbola corresponding to a different temperature. The temperature increases from $T_1$ to $T_5$.

Now, let's consider an ideal-gas isotherm, corresponding to constant $T$. We find, in this case, that

$$P = \frac{\text{constant}}{v}.$$

On a $P$–$v$ plot, this equation describes a hyperbola. We show five different hyperbolic ideal-gas isotherms in Figure 4.1, with temperature increasing from $T_1$ to $T_5$.

Notice that all of the ideal-gas isotherms are monotonic – that is, their slope has the same sign for all values of $v$. This means that the behavior of the system is uniform in nature. For example, if the ideal gas is compressed, so that $v$ is decreased, then the pressure $P$ of the gas increases. This is a characteristic of a stable system – specifically, if the piston in a cylinder is pushed inward, to decrease the volume, the pressure of the gas increases and acts to restore the system to its original state.

## Van der Waals' Modifications

In his PhD dissertation, van der Waals introduced two fairly simple modifications to the ideal-gas equation of state. These modifications address the major oversimplifications associated with the ideal gas.

### Modification One: Finite Molecular Volume

One of the main characteristics of the ideal-gas model is that the molecules are considered to be point particles – that is, particles of zero volume. In a real gas, the molecules have a finite volume. To include this aspect of real molecules, van der Waals introduced the quantity $b$ to represent the molar volume of the molecules in a gas. Thus, he made the following change to the ideal-gas model:

$$P = \frac{RT}{v} \rightarrow \frac{RT}{v - b}.$$

This change indicates that the smallest possible value for the molar volume $v$ is $b$. Note that $b$ has the units of cubic meters per mole, $m^3/mol$.

### Modification Two: Interparticle Attractions

The second oversimplification in the ideal gas is the omission of interactions between molecules. In a real gas, molecules experience an attractive force between one another. In the bulk of the gas this doesn't have much effect on the pressure because any given molecule feels attractions in all directions and they cancel out. Near the wall of the container, however, a molecule feels a net attractive force pulling it back toward the bulk of the gas, thus reducing the pressure.

This attractive force is a *pairwise* interaction between *neighboring pairs* of molecules. The probability that a molecule is found at a certain location is proportional to the molar density of the gas; that is, proportional to $n/V = 1/v$. Similarly, the probability that two molecules are next to one another is proportional to the molar density squared; that is, proportional to $1/v^2$. Thus, van der Waals reasoned that the attractive reduction in pressure should be of the form of $a/v^2$, where $a$ is a parameter determined by the type of molecules in the gas. Thus, we take the pressure to be reduced as

$$P \rightarrow P - \frac{a}{v^2}.$$

The units of the constant $a$ are $J \cdot m^3/mol^2$.

Combining these modifications, we can now write the complete *van der Waals equation of state* as follows:

$$P = \frac{RT}{v - b} - \frac{a}{v^2}. \tag{4.1}$$

Equivalently, stated in words, we have pressure exerted on container = thermal pressure – self-attraction.

The form of the van der Waals equation in Eq. (4.1) is the one we will use for the $P$–$v$ plots later in this chapter.

Another way to express the van der Waals equation of state is to move the self-attraction term, $-a/v^2$, to the left side of the equation, and then multiply both sides by $(v - b)$. This yields

$$\left(P + \frac{a}{v^2}\right)(v - b) = RT.$$

Equivalently, showing the dependence on the number of moles, $n$, we have

$$\left(P + \frac{an^2}{V^2}\right)(V - nb) = nRT.$$

This is the van der Waals modification of the corresponding ideal-gas expression

$$PV = nRT.$$

This version of the van der Waals equation is displayed on the wall of a building in Leiden, Netherlands, where van der Waals wrote his PhD dissertation at the University of Leiden. It is a common alternative expression that is found in many treatments of this topic.

With the appropriate values of $a$ and $b$, the van der Waals equation provides semiquantitative agreement with experiments on real gases. In Table 4.1 we list the values of $a$ and $b$ for a number of different substances.

As an example, the van der Waals equation for water is given by the expression

$$P_{\text{water}} = \frac{(8.314\ \text{J}/(\text{mol}\cdot\text{K}))T}{\left(v - 3.049 \times 10^{-5}\ \text{m}^3/\text{mol}\right)} - \frac{0.5523\ \text{J}\cdot\text{m}^3/\text{mol}^2}{v^2}. \tag{4.2}$$

We show $P$-versus-$v$ isotherms for this expression over a range of temperatures, from $T_1 = 560$ K to $T_5 = 800$ K, in Figure 4.2. Note that the van der Waals isotherms are quite different from the simple hyperbolas of the ideal-gas

Table 4.1 *The van der Waals constants* a *and* b *for various substances*

| Substance | $a$ ($\text{J}\cdot\text{m}^3/\text{mol}^2$) | $b$ ($\text{m}^3/\text{mol}$) |
|---|---|---|
| Ammonia | 0.4225 | $3.713 \times 10^{-5}$ |
| Argon | 0.1355 | $3.201 \times 10^{-5}$ |
| Helium | 0.00346 | $2.380 \times 10^{-5}$ |
| Hydrogen | 0.02453 | $2.651 \times 10^{-5}$ |
| Water | 0.5523 | $3.049 \times 10^{-5}$ |
| Nitrogen | 0.1370 | $3.870 \times 10^{-5}$ |
| Oxygen | 0.1382 | $3.186 \times 10^{-5}$ |

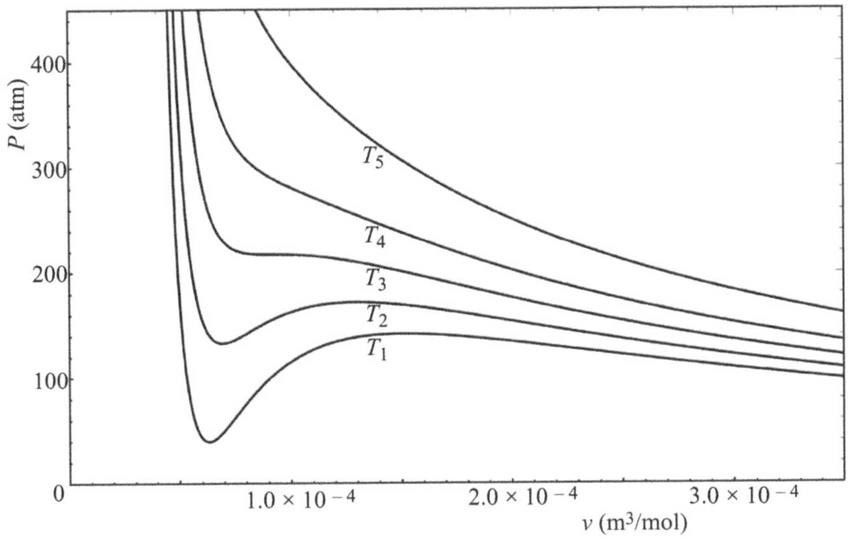


Figure 4.2 The van der Waals isotherms for water, with $a = 0.5523\ \mathrm{J \cdot m^3/mol^2}$, and $b = 3.049 \times 10^{-5}\ \mathrm{m^3/mol}$. The isotherms are for the following temperatures: $T_1 = 560\,\mathrm{K}$, $T_2 = 600\,\mathrm{K}$, $T_3 = 646\,\mathrm{K}$, $T_4 = 700\,\mathrm{K}$, $T_5 = 800\,\mathrm{K}$.

model; in fact, they show a range of different shapes. We turn now to the physical interpretation of these isotherms.

## Interpreting the van der Waals Isotherms

The first observation we make on looking at the isotherms for water in Figure 4.2 is that the high-temperature isotherms, $T_4$ and $T_5$, are similar in shape to the isotherms of the ideal gas. In particular, they are monotonic and not too different from the ideal-gas hyperbolas. This makes sense because at high temperature a gas behaves much like an ideal gas, with interactions between molecules that are small compared to their high thermal energies.

Next, we note that as the temperature is lowered, the isotherms take on more interesting shapes, as shown by the flat plateau for $T_3$, and the local maxima and minima that are seen for $T_2$ and $T_1$. This is a significant qualitative departure from the behavior seen in the ideal gas.

To understand the significance of the shape of the lower-temperature van der Waals isotherms, let's focus on the isotherm for $T_1$, as shown in Figure 4.3. First, note that the high-$v$ portion of the isotherm – to the right of point 2 – is simple and monotonic, much like the isotherm of an ideal gas. We associate

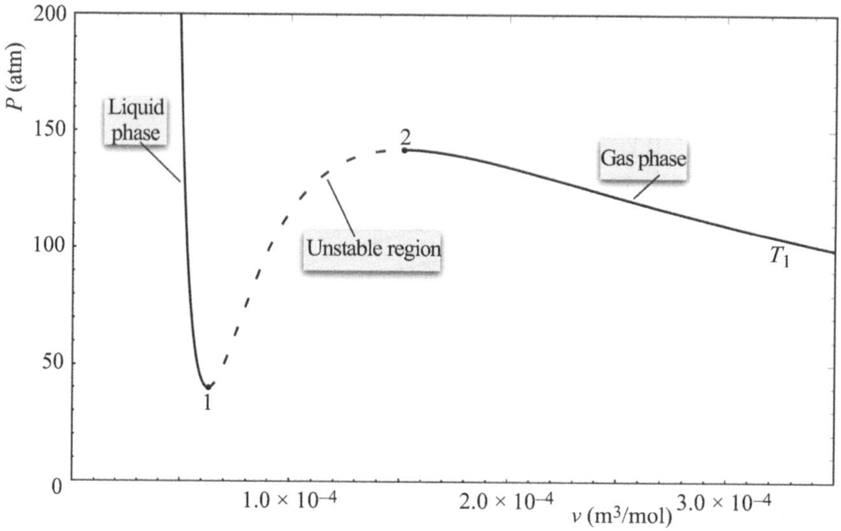


Figure 4.3 The van der Waals isotherms for water, with $a = 0.5523\ \text{J} \cdot \text{m}^3/\text{mol}^2$, and $b = 3.049 \times 10^{-5}\ \text{m}^3/\text{mol}$, and $T_1 = 560\ \text{K}$. The high-$v$ portion of the isotherm corresponds to the gas phase, and the low-$v$ portion corresponds to the liquid phase. In between these phases is an unstable region of the isotherm.

this portion of the isotherm with the gas phase, as indicated in the figure. In particular, we note that the isotherm implies that the gas is easily compressed – that is, it takes a large reduction in $v$ to produce a significant increase in the pressure, $P$. This is as expected for a gas.

On the other hand, note that on the low-$v$ portion of the curve – to the left of point 1 – a small reduction in $v$ results in a sharp increase in the pressure. This means that the system is practically incompressible along this branch of the isotherm, as expected for a liquid. Thus, we associate the low-$v$ portion of the isotherm with the liquid phase, as indicated in the figure.

Between the gas and liquid phases in Figure 4.3 we see a region that is labeled "unstable." To see the reason for this, note that between point 1 and point 2 the slope of the isotherm is positive. This means, for example, that an increase in molar volume $v$ causes an increase in pressure $P$. This may sound innocuous enough, but it has serious consequences.

For example, consider a cylinder filled with gas and held in place by a piston. Now, in the "unstable" region of the isotherm, suppose the piston is moved to the right, causing the volume of the gas to increase. As a result, the pressure of the gas increases. This means that the pressure pushes the piston farther to the right, increasing the volume of the gas more, which causes the pressure to increase more, which increases the volume even more, and so on. This is a runaway instability, and cannot represent a physical system in equilibrium. On the other hand, in the liquid and gas phases an increase in volume causes a decrease in pressure, which in turn tends to bring the piston back to its original position. Thus, both the liquid and gas are stable phases.

The question now is, how do we connect the liquid and gas phases on the two stable branches of the isotherm? Well, we know that when a gas condenses to a liquid the temperature and pressure are the same in the two coexisting phases. The temperature condition is taken care of because we are looking at a single isotherm. But what about the pressure? Apparently, we need a horizontal, constant-pressure line to connect the two phases. The value of this pressure, somewhere between the values associated with points 1 and 2, is determined by a procedure presented to the British Chemical Society by James Clerk Maxwell in 1875. We explore Maxwell's contribution in the next section.

## 4.2 The Maxwell Construction

Based on the discussion in Section 4.1, we expect a "physical" van der Waals isotherm to look like the solid curves and solid horizontal line shown in Figure 4.4. Here we see the liquid and gas phases connected by a line of constant pressure, extending from point B to point F. This represents the constant-temperature and constant-pressure coexistence between the liquid and gas phases. The dashed portion of the original isotherm, from C to E, is the *unstable* region, as mentioned previously. The dot-dashed portions of the original isotherm, from B to C and from E to F, are *metastable* regions. These will be discussed in more detail later in this section.

We turn now to the basic mechanism for connecting the liquid and gas phases, as proposed by Maxwell.

### The Gibbs Free Energy

Recall the following basic principles of equilibrium phases and phase transitions, as presented in Section 1.3. First, the Gibbs free energy is a minimum in an equilibrium phase. Second, when the minima in the Gibbs free energy of two different phases cross one another, as in Figure 1.7, a phase transition occurs as the minimum free energy shifts from one phase to the other.

To investigate the Gibbs free energy, $G$, for the case of a van der Waals isotherm, we start with the differential form, $dG$, as derived in Eq. (1.4):

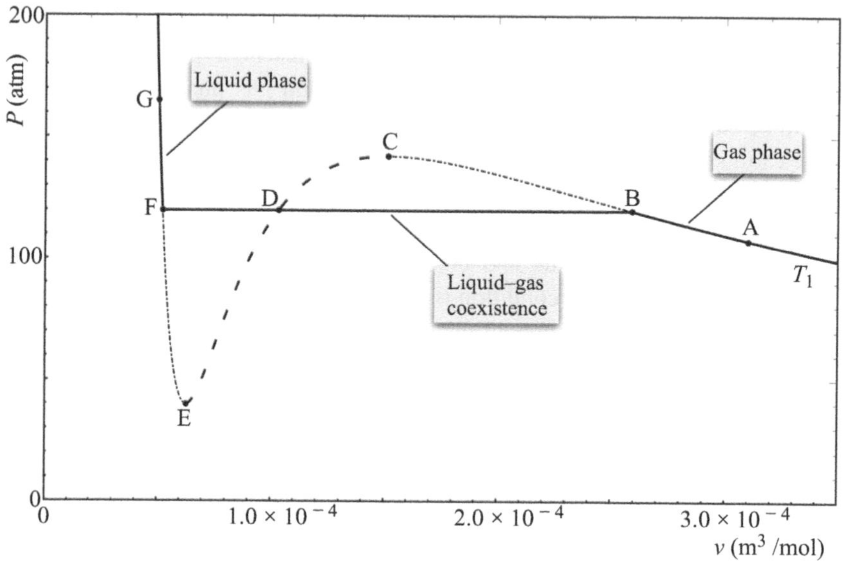


Figure 4.4 The "physical" van der Waals isotherm for water with $T_1 = 560$ K. Note the constant-pressure line connecting point B in the gas phase to point F in the liquid phase. The value of the pressure on this line is determined by equating the area of the region BCDB to the area of region DEFD.

$$dG = -SdT + VdP.$$

In our case we're investigating an isotherm, and hence the temperature is constant so $dT = 0$. It follows, then, that the change in the Gibbs free energy is simply

$$dG = VdP.$$

If we let $g = G/n$ be the *molar Gibbs free energy*, sometimes referred to as the *chemical potential*, $\mu$, and let $v$ be the molar volume, we can write

$$dg = vdP.$$

Integrating this expression from an arbitrary point 1 to a different point 2 gives the corresponding change in the molar Gibbs free energy:

$$g_2 - g_1 = \int_1^2 v\, dP. \tag{4.3}$$

We will use this relationship extensively in what follows.

To carry out an integral like this, we want $v$ to be the ordinate of our plot and $P$ to be the abscissa. Thus, we re-plot Figure 4.4 with $v$ on the $y$-axis and $P$ on

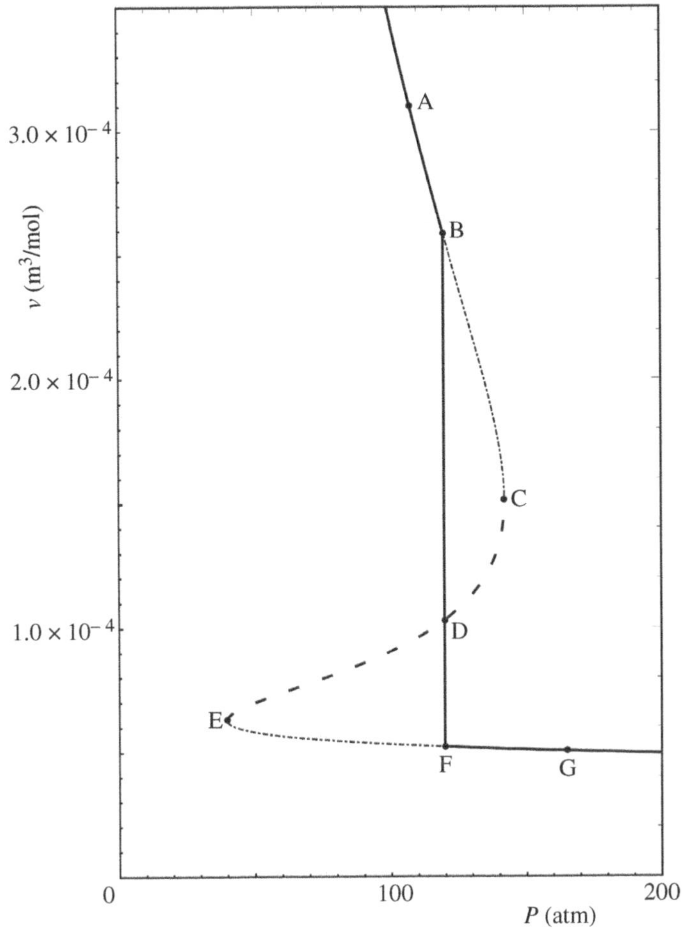


Figure 4.5 A re-plot of Figure 4.4, with $v$ on the $y$-axis and $P$ on the $x$-axis. Note that $v$ is a single-valued function of $P$ on the following segments: ABC, CDE, and EFG.

the $x$-axis. This is shown in Figure 4.5. Integrating under this curve gives the desired change in molar Gibbs free energy. Note, however, that $v$ is a multi-valued function of $P$, and thus we must consider separately the integral along each of the single-valued segments ABC, CDE, and EFG.

The integral along ABC gives a large positive value, since $v$ is large on this segment. Thus $g$ increases rapidly as we go from A to B to C. This is shown in Figure 4.6. From C to D to E, on the other hand, the integral in Eq. (4.3) is going in the negative direction, with $P$ decreasing, and hence $g$ decreases along

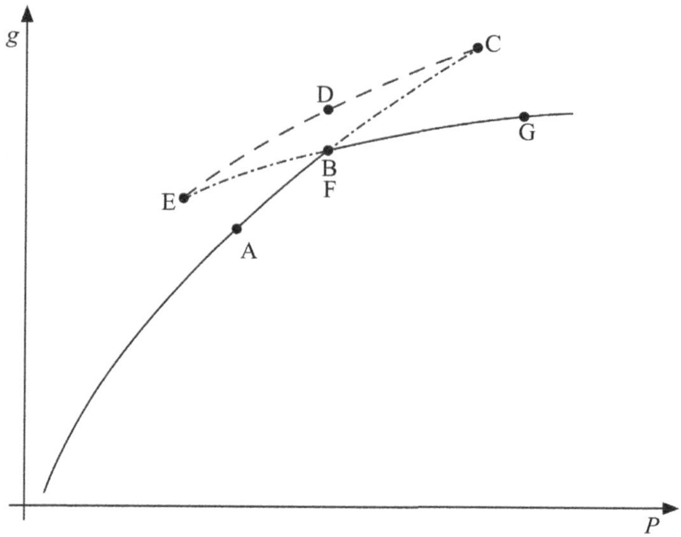


Figure 4.6 The molar Gibbs free energy for the isotherm shown in Figure 4.5. The crossover point, labeled with B and F, is the location of the liquid–gas phase transition.

this segment. This is also shown in Figure 4.6. Finally, from E to F to G the integral again gives a positive value, but this time with a relatively small magnitude. This is shown by the curve with the smallest upward slope in Figure 4.6.

Notice that the curves for $g$ cross at the point labeled with both B and F. At this point the global minimum of the system switches from the curve AB to the curve FG – this is the location of the liquid–gas phase transition. Recalling that $v = dg/dP$, we see that the molar volume is discontinuous at the phase transition, as illustrated by points B and F in Figure 4.6, because $g$ changes slope from the segment AB to the segment FG. The point D, which is at the same pressure as points B and F, is in the unstable region (dashed curve) and has a larger value of the free energy.

The dot-dashed segments in Figures 4.5 and 4.6, from B to C and from E to F, are *metastable* regions. As mentioned in Chapter 1, this means that the minimum of $g$ on these segments is a local minimum, rather than the global minimum corresponding to the equilibrium phase.

To understand this more completely, imagine picking up a ball from the floor, holding it in your hand, and raising it to chest level. In this case, the global minimum of the gravitational potential energy is the floor, but the ball is in a local minimum formed by the cupped shape of your hand. As you raise the ball, it remains in the local minimum of your hand, even though it is now well above the global minimum of the system. This is like moving beyond the segment AB onto the segment BC, with the system remaining in the local minimum of one phase, even though the global minimum is in a different phase.

If someone comes along now and bumps into your arm, the ball can bounce out of the local minimum in your hand and fall rapidly to the global minimum on the floor. Thus, a jostle to the system, or a nucleation site, can cause the system to collapse rapidly to the global minimum corresponding to the other phase. Superheated and supercooled systems are examples of this, where the system remains in its original phase until a jostle of some sort causes a sudden phase transition to the true equilibrium phase.

## The Equal-Area Condition

When we look carefully at Figure 4.6, we see that $g$ increases from B to C, decreases from C to D and from D to E, and then increases again from E to F. The net change in $g$ from B to F is zero, of course, because the finishing point F has the same value of $g$ as the beginning point B. Thus, zero change in $g$ is the condition that identifies the location of the phase transition.

We can express this in terms of integrals of $vdP$. For example, the condition for the liquid–gas phase transition can be stated as

$$\int_{\text{Point B}}^{\text{Point F}} v\, dP = 0.$$

Let's write this out explicitly, in terms of integrals on each separate segment of the isotherm. We find the following:

$$\int_{P_{\mathrm{B}}}^{P_{\mathrm{C}}} v_{\mathrm{BC}}\, dP + \int_{P_{\mathrm{C}}}^{P_{\mathrm{D}}} v_{\mathrm{CD}}\, dP + \int_{P_{\mathrm{D}}}^{P_{\mathrm{E}}} v_{\mathrm{DE}}\, dP + \int_{P_{\mathrm{E}}}^{P_{\mathrm{F}}} v_{\mathrm{EF}}\, dP = 0.$$

This is the condition for the phase transition.

Now, the middle two integrals in this expression are negative because we're integrating in the direction of decreasing $P$ on those segments. Let's pull the negative sign out in front of these integrals and reverse the range of integration, so that no change is made to the overall expression. In this way, however, it's clear where the positive and negative contributions are being made. This gives

$$\int_{P_{\mathrm{B}}}^{P_{\mathrm{C}}} v_{\mathrm{BC}}\, dP - \int_{P_{\mathrm{D}}}^{P_{\mathrm{C}}} v_{\mathrm{CD}}\, dP - \int_{P_{\mathrm{E}}}^{P_{\mathrm{D}}} v_{\mathrm{DE}}\, dP + \int_{P_{\mathrm{E}}}^{P_{\mathrm{F}}} v_{\mathrm{EF}}\, dP = 0.$$

We now rearrange to put the first two integrals on one side of the equation and the second two integrals on the other side. This gives

$$\int_{P_B}^{P_C} v_{BC}\, dP - \int_{P_D}^{P_C} v_{CD}\, dP = \int_{P_E}^{P_D} v_{DE}\, dP - \int_{P_E}^{P_F} v_{EF}\, dP.$$

Finally, we note that the pressure range from D to C is the same as the pressure range from B to C. Similarly, the pressure range from E to F is the same as the pressure range from E to D. Thus, we can rewrite the integrals as follows:

$$\int_{P_B}^{P_C} v_{BC}\, dP - \int_{P_B}^{P_C} v_{CD}\, dP = \int_{P_E}^{P_D} v_{DE}\, dP - \int_{P_E}^{P_D} v_{EF}\, dP.$$

Combining the integrals with the same pressure ranges yields

$$\int_{P_B}^{P_C} (v_{BC} - v_{CD})\, dP = \int_{P_E}^{P_D} (v_{DE} - v_{EF})\, dP.$$

The advantage of writing the phase-transition condition in this way is that it has a very useful geometric interpretation.

To see this, note that the integrand in the left-hand integral, $(v_{BC} - v_{CD})$, is the vertical distance between the segments BC and CD. Integrating this horizontally from $P_B$ to $P_C$ yields the area of the region bounded by BCDB, and labeled Area 1 in Figure 4.7. Similarly, the integrand in the right-hand integral, $(v_{DE} - v_{EF})$, is the vertical distance from DE to EF. Integrating this horizontally from $P_E$ to $P_D$ yields the area of the region bounded by DEFD, and labeled Area 2 in Figure 4.7. Thus, the condition for the phase transition is that these areas must be equal:

$$\text{Area 1} = \text{Area 2}.$$

This is referred to as the *Maxwell construction*.

We now revert to the original way of plotting the isotherm, with $P$ vertical and $v$ horizontal. This gives the result shown in Figure 4.8. Notice that the level line at constant pressure is drawn at the location where the area above the line to the isotherm, Area 1, is equal to the area below the line to the isotherm, Area 2. This provides a convenient way of visualizing the Maxwell condition.

To determine the coexistence pressure numerically, we use the following procedure. First, choose a value for the coexistence pressure, $P_{coex}$, in the range from $P_E$ to $P_C$. Next, use the expression for $P_{water}$ from Eq. (4.2) to determine the molar volumes corresponding to this pressure. We find three solutions, $v_F$, $v_D$, and $v_B$. Use the smallest and largest of these molar volumes, $v_F$ and $v_B$, respectively, to carry out the following integral:

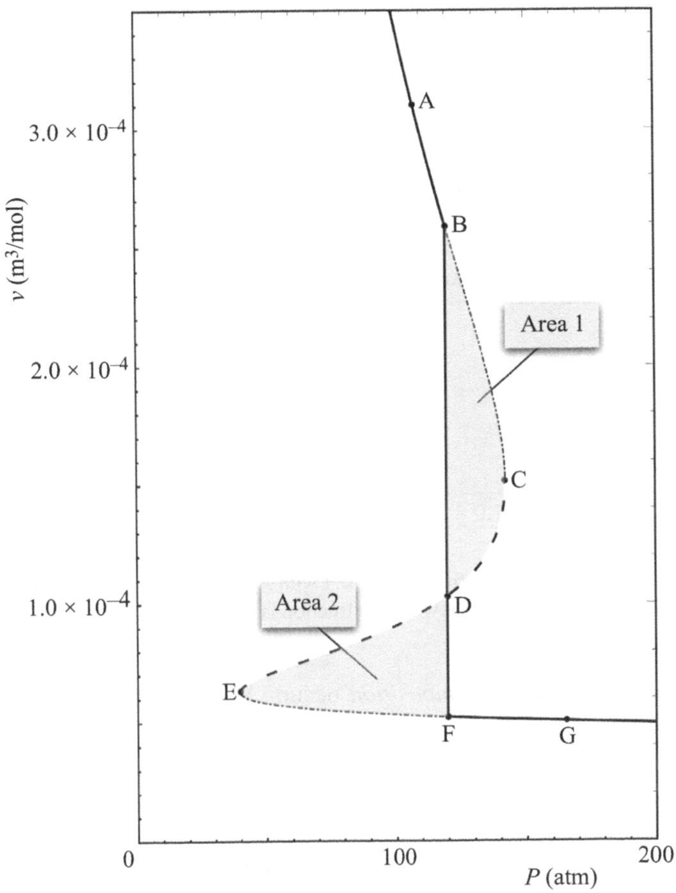


Figure 4.7 The isotherm from Figure 4.5 with the relevant areas, Area 1 and Area 2, shaded. The Maxwell construction is based on the condition that Area 1 = Area 2.

$$\int_{v_F}^{v_B} (P_{\text{water}} - P_{\text{coex}})dv = 0.$$

If this integral yields a positive value, which means that Area 1 is greater than Area 2, choose a $P_{\text{coex}}$ that is a bit larger and repeat; if the integral is negative, because Area 2 is greater than Area 1, choose a smaller value for $P_{\text{coex}}$ and repeat. Continue with this process until the integral is equal to zero within the desired accuracy.

Next, we recall that the dot-dashed curve from B to C is the metastable extension of the gas phase. Point C is the terminus of this metastable phase, beyond which the system is completely unstable. We refer to point C as the

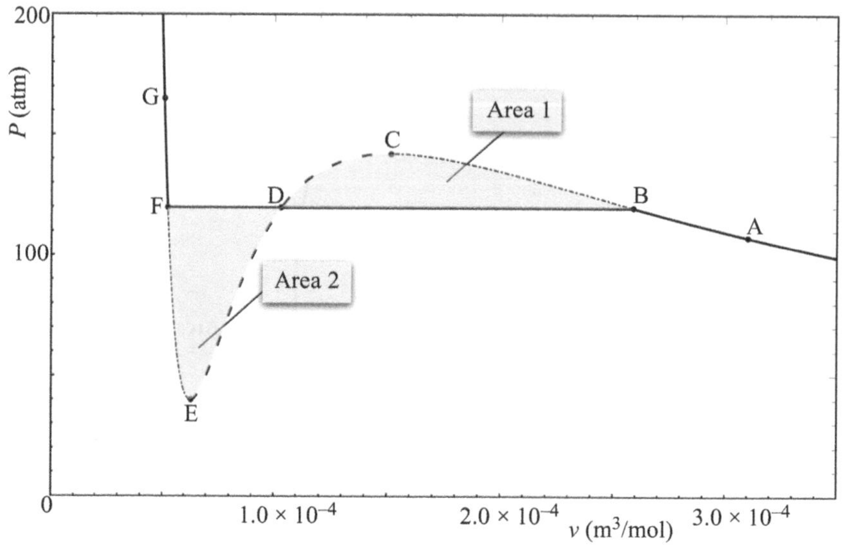


Figure 4.8 The isotherm from Figure 4.4 with Area 1 and Area 2 highlighted with shading. As before, the Maxwell construction is based on the condition Area 1 = Area 2.

point where *spinodal decomposition* occurs. What this means is that a phase transition from gas to liquid occurs within the entire body of the gas phase at this point, driven by diffusion rather than occurring at nucleation sites. This mechanism results in a periodic, wave-like variation of properties such as density throughout the system, giving rise to the name spinodal. Similarly, point E is the point of spinodal decomposition of the liquid phase.

Finally, in Figure 4.9 we show the van der Waals isotherms for water from Figure 4.2, this time including the constant-pressure lines determined with the Maxwell construction. These lines show the coexistence of the liquid and gas phases for the temperatures $T_1$ and $T_2$.

Notice that the range in values of $v$ for coexistence diminishes with increasing temperature, and eventually vanishes altogether at the temperature $T_3$. We explore the coexisting phases in the next section, and this special temperature in Section 4.4.

## 4.3 Coexisting Phases and the Lever Rule

We turn now to the coexistence of phases associated with the van der Waals equation of state. We also examine the relative composition of the two phases in the coexistence region.

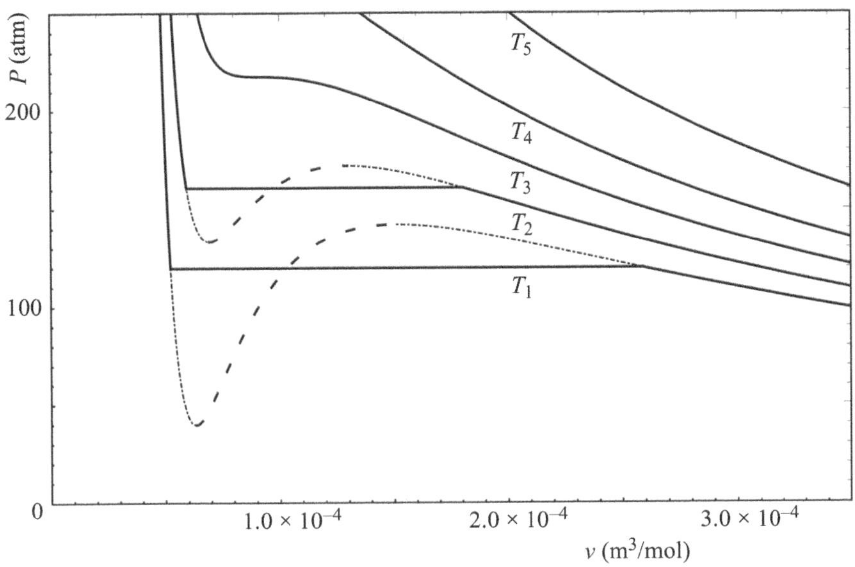


Figure 4.9 The van der Waals isotherms for water with the Maxwell construction shown for the temperatures $T_1$ and $T_2$. The coexistence regions shown for $T_1$ and $T_2$ vanish at the temperature $T_3$.

## Coexisting Phases

We've seen how to determine the coexisting molar volumes for a given isotherm by applying the equal-area condition. If we carry out this procedure over a range of temperatures for water, we obtain the black line shown in Figure 4.10. For context, the underlying isotherms are shown in gray.

Notice the dome shape of the coexistence curve, which is characteristic for all substances. In addition, we see that the molar volume is rather large on the gas phase side of the curve, and much smaller on the liquid side. This leads to an asymmetry of the coexistence curve.

To understand the coexistence curve more fully, imagine starting with a system of gaseous water (steam) at the temperature $T_1 = 560$ K in Figure 4.10, and a fairly large molar volume of 300 $\text{cm}^3/\text{mol}$. Now, we compress the system, keeping the temperature the same. At the point shown by a black dot in the figure at about 260 $\text{cm}^3/\text{mol}$ the gas begins to condense to liquid water. As the molar volume is reduced further, to 200 $\text{cm}^3/\text{mol}$ and 100 $\text{cm}^3/\text{mol}$, more and more gas condenses and the amount of liquid increases. Finally, at a molar volume of about 50 $\text{cm}^3/\text{mol}$, the system is completely liquid and further compression results in a rapid increase in pressure.

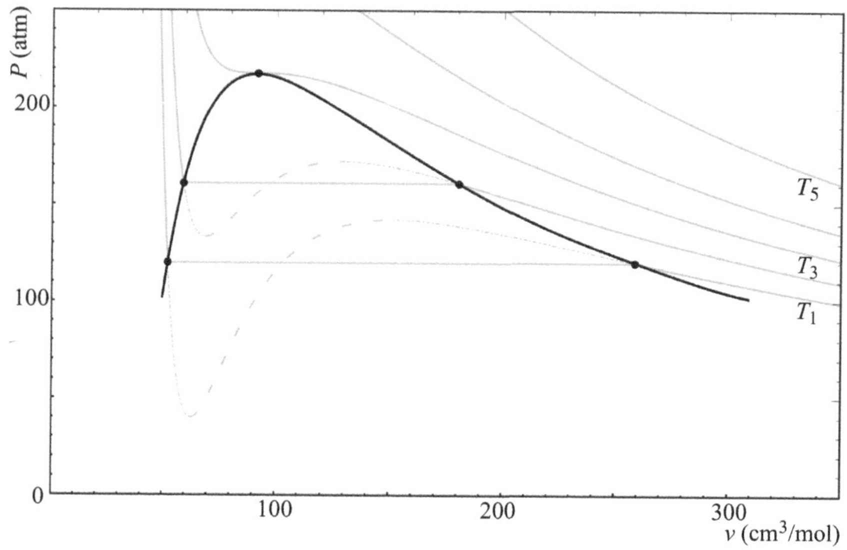


Figure 4.10 The coexistence regions (black line) for the van der Waals equation of state for water. The gray isotherms are the same as those shown in bold in Figure 4.9.

We now show how to determine the specific fraction of gas and liquid that coexist for any given value of the molar volume.

## The Lever Rule

To determine the fraction of gas and liquid that coexist at a given molar volume, $v$, we refer to Figure 4.11. First, we designate the fraction of the system that is gas by $f_g$, and the fraction that is liquid by $f_l$. The system contains only gas and liquid, and hence it follows that these fractions must add to one:

$$f_l + f_g = 1. \tag{4.4}$$

Second, we note in Figure 4.11 that the molar volume of the gas phase is $v_g$, and the molar volume of the liquid phase is $v_l$. Combining these fractions and molar volumes, we can write

$$f_l v_l + f_g v_g = v. \tag{4.5}$$

The two conditions expressed by Eqs. (4.4) and (4.5) determine the two fractions in terms of $v_g$ and $v_l$.

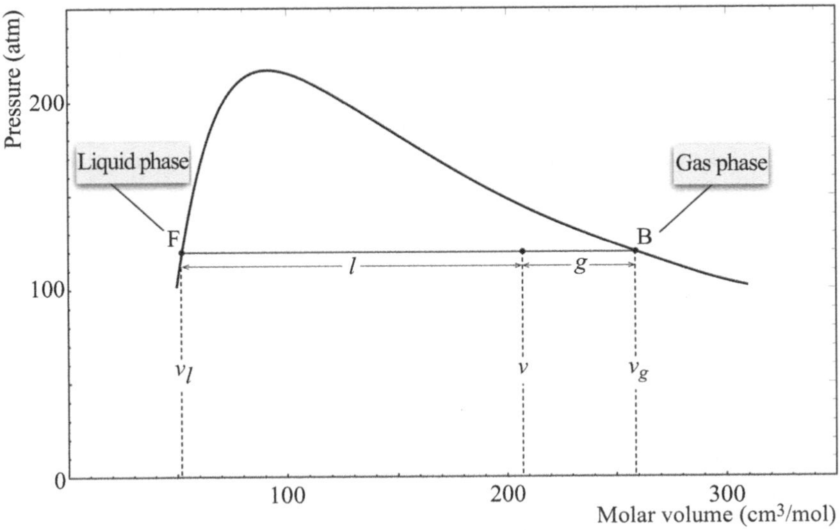


Figure 4.11 The lever rule determines the fraction of liquid and gas in coexistence when the molar volume of the system is $v$.

For example, we can use Eq. (4.4) to replace $f_l$ in terms of $f_g$; that is, $f_l = (1 - f_g)$. Substituting into Eq. (4.5), we find

$$(1 - f_g)v_l + f_g v_g = v.$$

Solving for $f_g$ yields

$$f_g = \frac{v - v_l}{v_g - v_l}.$$

The fraction on the right-hand side of this equation has interesting geometric interpretations.

To see this, we start with the numerator. Note in Figure 4.11 that $v - v_l$ is the "molar volume distance" from $v$ to the liquid phase at point F. This distance is labeled $l$ in the figure. Similarly, $v_g - v_l$ in the denominator is the molar volume distance from the gas phase at point B to the liquid phase at point F. Using the labels in the figure, this distance is $l + g$. Combining these results, we have

$$f_g = \frac{l}{l + g}. \tag{4.6}$$

Similarly, for the fraction of the system in the liquid phase we have

$$f_l = \frac{g}{l + g}. \tag{4.7}$$

Expressed this way, we see that the fraction of the system in the gas phase is proportional to the distance to the liquid phase, and the fraction in the liquid phase is proportional to the distance to the gas phase.

Note that the expressions for both fractions have the same denominator, $l+g$. It follows that the ratio of the fractions eliminates the denominator, and yields the following simple result:

$$\frac{f_g}{f_l} = \frac{l}{g}.$$

Rearranging, we find that

$$gf_g = lf_l.$$

This relationship is referred to as the *lever rule*, because the "distance" times the fraction for the gas phase is equal to the "distance" times the fraction for the liquid phase, just like in a lever where the distance times the force on one side of the lever is equal to the distance times the force on the other side.

The fractions given in Eqs. (4.6) and (4.7) can be rewritten in such a way that the fraction of the gas phase depends specifically on the distance to the gas phase, and the fraction of the liquid phase depends specifically on the distance to the liquid phase. To see how, let's start with Eq. (4.6) for the fraction of the gas phase:

$$f_g = \frac{l}{l+g}.$$

Now, let's add and subtract $g$ in the numerator. This yields

$$f_g = \frac{l+g-g}{l+g}.$$

Simplifying this expression gives

$$f_g = 1 - \frac{g}{l+g}. \tag{4.8}$$

This expression depends directly on the distance to the gas phase. Similarly, for the liquid fraction we have

$$f_l = 1 - \frac{l}{l+g}. \tag{4.9}$$

Equations (4.6) and (4.7) and Eqs. (4.8) and (4.9) are two distinct ways to express the same results.

As an example, in Figure 4.11 we have a liquid distance that is three times greater than the gas distance; that is, $l = 3g$. It follows that the fractions of the phases are as follows:

$$f_g = 1 - \frac{1}{4} = \frac{3}{4},$$
$$f_l = 1 - \frac{3}{4} = \frac{1}{4}.$$

So when the molar volume of the system is closer to the gas phase, as in Figure 4.11, there is more gas in the system, and when it is closer to the liquid phase there is more liquid in the system.

## 4.4 The Critical Point and Corresponding States

In this section we look at a special point of the van der Waals equation of state, referred to as the critical point. We then use this critical point to produce a reduced equation of state that applies to *all* liquid–gas systems.

### The Critical Point

In Figure 4.12 we show the gas–liquid coexistence curve for water. This is the same bold, dome-shaped curve shown in Figure 4.10, but without the gray isotherms in the background. Note the horizontal, constant-pressure regions of coexistence. The dashed horizontal lines are regions where the system is unstable, and the dot-dashed lines are the metastable regions. Notice that these regions diminish in size as one moves upward in pressure, eventually shrinking to a single point at the top of the coexistence curve. This single point is referred to as the *critical point*. It is here that the distinction between the gas and liquid phases vanishes. (We consider the precise shape of the coexistence curve near the critical point in Chapter 5, when we discuss critical exponents.)

We now derive the mathematical conditions that define the critical point. There are two such conditions, as indicated in Figure 4.13. For the first one, note that the isotherm corresponding to the critical temperature – the critical isotherm – is horizontal at the point where it touches the critical point. Therefore, the first condition is that the derivative of $P$ with respect to $v$ is zero at the critical point; that is, $\frac{dP}{dv} = 0$.

For the second condition, note that the location of the maximum and minimum of a low-temperature isotherm – at the junctures between the unstable and metastable regions, as shown by the lower two black dots in Figure 4.13 – also correspond to $\frac{dP}{dv} = 0$. In addition, the second derivative of the isotherm with respect to $v$ is negative at the maximum, and is positive at the minimum. As we move upward in temperature toward the critical point, the

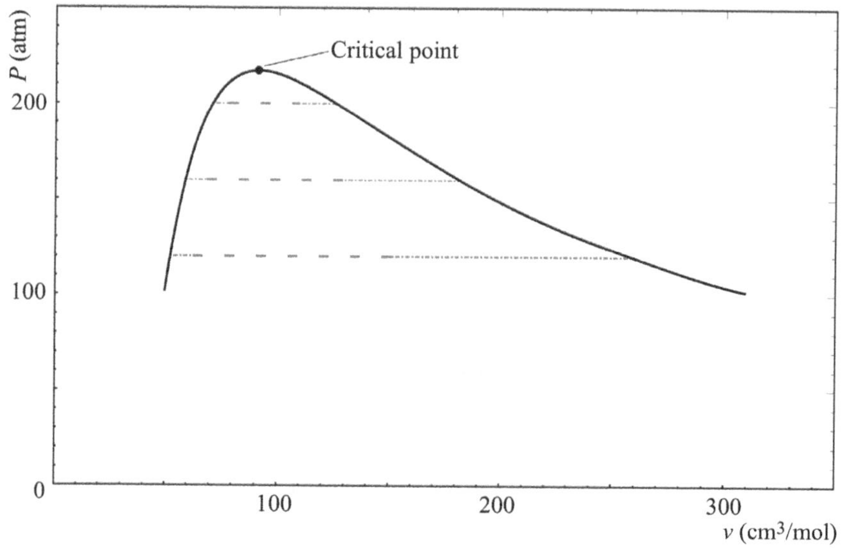


Figure 4.12 The dome-shaped coexistence curve for water. The horizontal lines indicate regions of coexistence between the gas and liquid phases. Specifically, the dashed lines indicate unstable regions, and the dot-dashed lines indicate metastable regions. The single point at the top of the coexistence dome is the critical point of the system.

maximum and minimum merge together into a single point. As a result, the second derivative – which can't be both finite positive and finite negative at the same time – must vanish there, so $\frac{d^2P}{dv^2} = 0$. Thus, the critical isotherm is defined mathematically as an inflection point in the isotherm, which means that both the first and second derivatives are equal to zero there.

To implement these conditions, we return to the expression for the van der Waals isotherm:

$$P = \frac{RT}{v-b} - \frac{a}{v^2}.$$

We now calculate the derivatives and set them to zero to determine the location of the critical point.

#### First Condition for the Critical Point: The First Derivative Is Zero

For the first condition, we take the derivative of $P$ with respect to $v$ and set it equal to zero:

$$\frac{dP}{dv} = -\frac{RT}{(v-b)^2} + \frac{2a}{v^3} = 0.$$

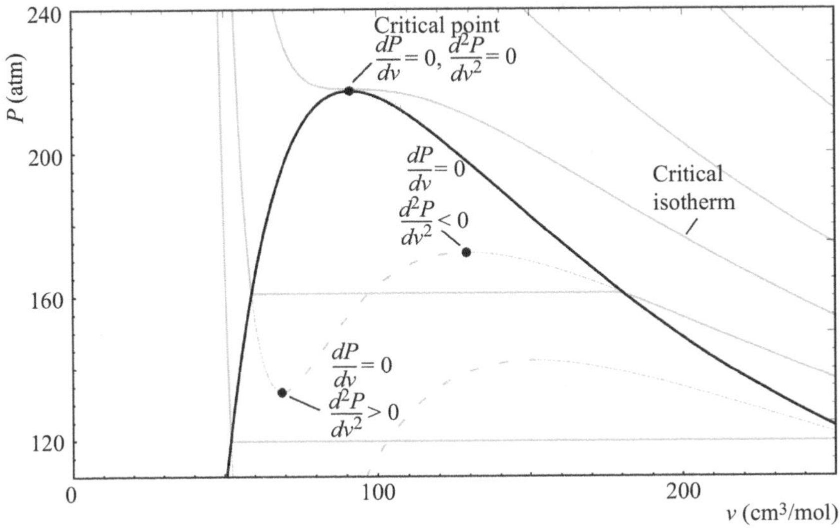


Figure 4.13 Zoomed in view of the dome-shaped coexistence curve for water. Low-temperature isotherms intersect the coexistence curve at two points connected by a horizontal line; the critical isotherm touches the coexistence curve at a single point – the critical point. Note that both the first and second derivatives of $P$ with respect to $v$ are zero at the critical point.

Evaluating this derivative for the critical temperature, $T_c$, and critical molar volume, $v_c$, yields

$$\frac{RT_c}{(v_c - b)^2} = \frac{2a}{{v_c}^3}.$$

Solving for $T_c$ in terms of $v_c$ gives

$$T_c = \frac{2a}{R{v_c}^3}(v_c - b)^2. \quad (4.10)$$

We will incorporate this result with the second condition for the critical point.

### Second Condition for the Critical Point: The Second Derivative Is Zero

We now take the second derivative of $P$ with respect to $v$ and set it equal to zero as well. The result is as follows:

$$\frac{d^2P}{dv^2} = \frac{2RT}{(v-b)^3} - \frac{6a}{v^4} = 0.$$

Substituting our result for $T_c$ from Eq. (4.10), we find

$$\frac{2R}{(v_c - b)^3}\left[\frac{2a}{Rv_c{}^3}(v_c - b)^2\right] = \frac{6a}{v_c{}^4}.$$

Simplifying and rearranging yields

$$\frac{2}{(v_c - b)} = \frac{3}{v_c}.$$

Solving this relation for the critical molar volume gives

$$v_c = 3b.$$

Substituting this result into Eq. (4.10) for $T_c$ we find that

$$T_c = \frac{2a}{R(3b)^3}(2b)^2 = \frac{8a}{27Rb}.$$

Finally, putting both $T_c$ and $v_c$ into the original van der Waals equation gives the critical pressure, $P_c$:

$$P_c = \frac{R8a}{27Rb(2b)} - \frac{a}{9b^2} = \frac{4a}{27b^2} - \frac{a}{9b^2} = \frac{a}{27b^2}.$$

Thus, the conditions that the first and second derivative are equal to zero yield the critical properties, and these are totally defined by the van der Waals constants, $a$ and $b$.

As an example, for water we find from Table 4.1 that $a = 0.5523\ \text{J}\cdot\text{m}^3/\text{mol}^2$ and $b = 3.049 \times 10^{-5}\ \text{m}^3/\text{mol}$. It follows that the critical molar volume is $v_c = 9.15 \times 10^{-5}\ \text{m}^3/\text{mol}$, the critical temperature is $T_c = 646\ \text{K}$, and the critical pressure is $P_c = 217$ atm. Note that these are the values of $P$ and $v$ that are seen for the critical point in Figure 4.13.

## Corresponding States

Now that we've defined the critical properties of the van der Waals equation, it's useful to write a "reduced" van der Waals equation, defined by dividing all quantities by their critical value. Denoting reduced quantities with the subscript "r," we make the following definitions:

$$P_r = P/P_c,$$
$$T_r = T/T_c,$$
$$v_r = v/v_c.$$

We will use these quantities to define a new, reduced equation of state.

Let's return to the original van der Waals equation,

$$P = \frac{RT}{v-b} - \frac{a}{v^2}.$$

We now replace $P$ with the following:

$$P = P_r P_c = P_r\left(\frac{a}{27b^2}\right).$$

Similarly, we have

$$v = v_r v_c = v_r(3b),$$

$$T = T_r T_c = T_r\left(\frac{8a}{27Rb}\right).$$

Making all of these substitutions gives

$$P_r\left(\frac{a}{27b^2}\right) = \frac{RT_r\left(\frac{8a}{27Rb}\right)}{v_r(3b)-b} - \frac{a}{v_r^2 9b^2}.$$

Simplifying, we find that

$$P_r = \frac{8T_r}{3v_r - 1} - \frac{3}{v_r^2}. \tag{4.11}$$

This is the reduced equation of state.

We plot the reduced equation of state in Figure 4.14. This equation *applies to the liquid and gas phases of all substances* as long as the reduced quantities of one substance are the same as the reduced quantities of another substance. This detailed connection between substances is referred to as the *law of corresponding states.*

As a specific example, note that Figure 4.14, which is plotted in terms of reduced quantities, is very similar to Figure 4.2, which is plotted in terms of absolute quantities for the case of water. In fact, the figures are actually the same. To see the connection, let's start with the lowest isotherm, which is labeled $T_1$ in both figures. In Figure 4.14 the dimensionless reduced temperature is $T_1 = 0.867$. In Figure 4.2 the lowest isotherm corresponds to $T_1 = 560\,\mathrm{K}$, but this is just equal to $(0.867)\,T_c$, where $T_c = 646\,\mathrm{K}$ for water. Similarly, $T_2$ in Figure 4.14 is $T_2 = 0.929$, and in Figure 4.2 it is $T_2 = (0.929)646\,\mathrm{K} = 600\,\mathrm{K}$. All of the other isotherms in these two figures are related in exactly the same way, each one connected by a given fraction of the critical temperature of water.

Thus, the law of corresponding states simply means that the reduced equation in Eq. (4.11) is a *universal equation of state* that applies to any substance, when

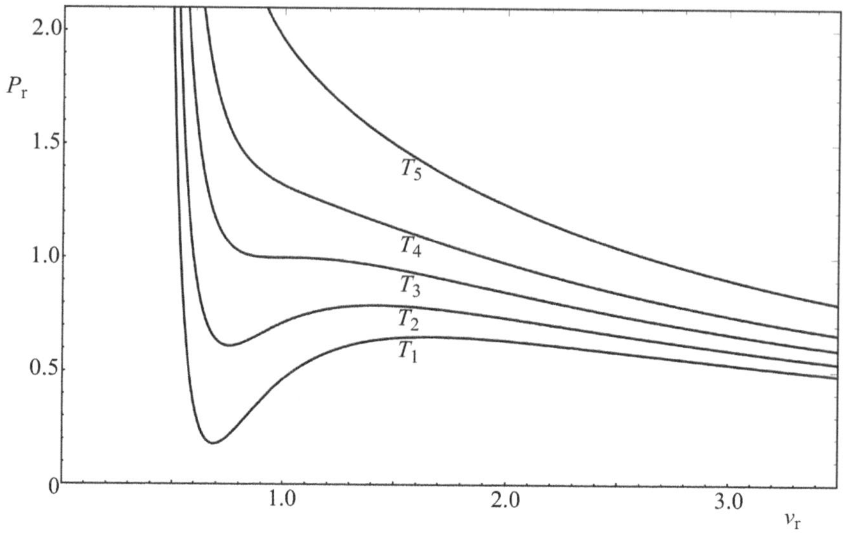


Figure 4.14 Reduced van der Waals isotherms. The reduced (dimensionless) temperatures for the isotherms are $T_1 = T_r = 0.867$, $T_2 = T_r = 0.929$, $T_3 = T_r = 1$, $T_4 = T_r = 1.08$, and $T_5 = T_r = 1.24$.

normalized to that substance's critical temperature. As another example, for ethane the critical temperature is $T_c = 300\,\text{K}$, and hence an ethane isotherm at $0.867T_c = 260\,\text{K}$ will look the same as the $T_1$ isotherm in Figure 4.14. Similarly, the isotherm for ethane at the temperature $1.1T_c = 324\,\text{K}$ is the same as $T_4$ in Figure 4.14. This illustrates why we say that the isotherms in Figure 4.14 are universal – valid for virtually any liquid–gas system – once you normalize with respect to the critical temperature of a particular substance.

## 4.5 Problems

4.1 **Isotherms.** Consider the van der Waals isotherm for water at 560 K. **(a)** What are the values of $v$ in cm$^3$/mol that give a pressure of 100 atm? **(b)** What is the interpretation of each of these solutions?

4.2 **Isotherms.** Consider the van der Waals isotherm for water at $T = 620\,\text{K}$. **(a)** Find the values of $P$ and $v$ at the spinodal point for the gas phase. **(b)** Find the values of $P$ and $v$ at the spinodal point for the liquid phase.

4.3 **Isotherms.** Consider the van der Waals isotherm for water at $T = 620\,\text{K}$. Find **(a)** the pressure at coexistence and **(b)** the values of the coexisting molar volumes for both the gas phase and the liquid phase.

4.4 **Critical properties.** Referring to Table 4.1, find **(a)** the critical temperature and **(b)** the critical pressure of ammonia.

4.5 **Critical properties.** Referring to Table 4.1, find **(a)** the critical temperature and **(b)** the critical pressure of nitrogen.

4.6 **Critical properties.** **(a)** Show that the product $Pv/T$ is a constant for an ideal gas. Determine the value of this constant.

**(b)** Show that the product of the critical properties $P_c v_c / T_c$ is a constant for a van der Waals substance as well. Determine the value of this constant.

**(c)** If the critical pressure and critical molar volume for nitrogen are $P_c = 3.39 \times 10^6$ J/m$^3$ and $v_c = 1.16 \times 10^{-4}$ m$^3$/mol, respectively, what is the critical temperature for nitrogen?

4.7 **Corresponding states.** The isotherm of oxygen at 167.4 K is the same as the isotherm labeled $T_4$ in Figure 4.14.

**(a)** What is the critical temperature of oxygen?

**(b)** At what temperature is the isotherm of oxygen the same as the isotherm labeled $T_2$ in Figure 4.14?

4.8 **Corresponding states.** The critical temperature and pressure of methane are $T_c = 190$ K and $P_c = 46$ atm, respectively. At what temperature is the isotherm of methane the same as the isotherm labeled **(a)** $T_1$ and **(b)** $T_4$ in Figure 4.14?

4.9 **Corresponding states.** The critical temperature and pressure of carbon dioxide are $T_c = 304$ K and $P_c = 73$ atm, respectively. At what temperature is the isotherm of carbon dioxide the same as the isotherm labeled **(a)** $T_2$ and **(b)** $T_5$ in Figure 4.14?

# 5

# Magnets and Mean-Field Theory

Up to this point, we have studied phase transitions in solids, liquids, and gases. Now we expand our attention to magnets and the phase transitions that occur in them. As we shall see, there are deep connections between the phase transitions in these seemingly disparate types of systems.

## 5.1 Magnetic Phase Transitions

When we think about magnets, we tend to imagine rods of metal, usually iron, that are magnetized such that they have a North-seeking magnetic pole on one end and a South-seeking magnetic pole on the other end. Such metallic rods are referred to as "bar magnets" or "permanent magnets." These are the familiar magnets found in schoolroom laboratories.

Though these magnets may seem *permanent* by everyday standards, the magnetism of a bar magnet can in fact change. For example, if the temperature of an iron rod is increased enough there is a phase transition, with the result that its magnetization disappears. The temperature at which the magnetization vanishes – the magnet's critical point – is referred to as the *Curie point.* In iron, the Curie point is 770 °C. Above this temperature an iron bar can no longer become a permanent magnet.

An excellent real-world example of a magnetic phase transition with temperature is provided by the phenomenon of seafloor spreading. As has been known since the 1950s, the ocean floor is characterized by structures known as *mid-ocean ridges.* Along these ridges, molten magma rises, moves outward in both directions away from the ridge, and cools. This process is illustrated in Figure 5.1.

When the magma is first extruded, it is at a temperature well above the Curie point, and hence it has no inherent magnetism of its own. The applied external

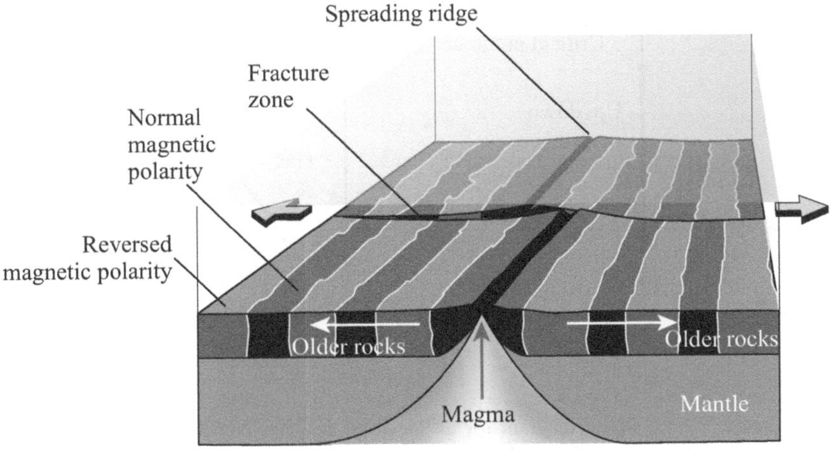


Figure 5.1 The process of seafloor spreading from a mid-ocean ridge. As the molten magma moves away from the ridge and cools, it locks in the *current* direction of Earth's magnetic field, resulting in paired strips of alternating magnetism on either side of the ridge as Earth's magnetic field periodically reverses. Adapted with permission, from Fossen H, Teyssier C. *Plate Tectonics*. Cambridge: Cambridge University Press, 2024.

magnetic field of Earth induces a magnetism in it, however. As the magma moves away from the ridge and cools, its temperature drops below the Curie point. This "freezes" or "locks" the induced magnetization in place.

All of this is well and good, but things really get interesting when you consider the fact that Earth's magnetic field is not constant, as one might think. Instead, Earth's magnetic field *flips direction* roughly every 300,000 years in a process known as a *geomagnetic reversal*. The result is that the magnetism on either side of the ridge alternates in direction in strips parallel to the ridge, as indicated by the arrows in Figure 5.1.

Finding these magnetic "signatures" in the seafloor provides conclusive evidence for plate tectonics and continental drift. Not only that, Earth seems to be long overdue for another reversal, and there are some indications that the process is beginning to happen. The cause of the reversals is due to the mag netohydrodynamics of Earth's fluid core, but this is not yet understood in all of its details.

To put the temperature behavior of magnets in proper context, we show a schematic phase diagram for a magnet in Figure 5.2(a). Here we plot the temperature $T$ on the vertical axis and the magnetic field $B$ on the horizontal axis. When the magnetic field is positive the magnet is in the up-magnetized phase, and when it is negative the system is in the down-magnetized phase.

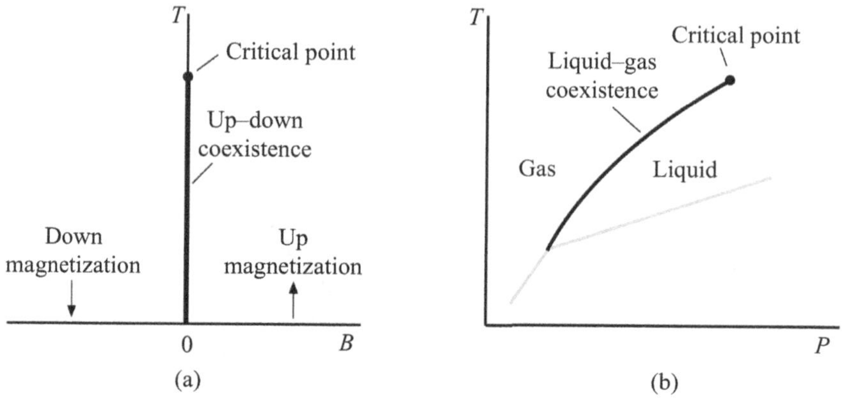


Figure 5.2 (a) Phase diagram for an idealized magnet. Notice the coexistence between the up and down phases until the temperature is raised to the critical point. (b) Phase diagram for an idealized liquid–gas system. Coexistence between the liquid and gas phases ends at the critical point.

Thus, at zero magnetic field and low temperature there is a coexistence between the up and down phases, as indicated by the bold vertical line. This coexistence ends for temperatures above the critical point.

The corresponding phase diagram for a liquid–gas system is shown in Figure 5.2(b). In this case, we plot temperature $T$ versus pressure $P$. Without the simple up–down symmetry of the magnet, the coexistence curve between liquid and gas is no longer a vertical straight line. The coexistence ends at the critical point, however, just as it does for the magnet.

A different way to look at these phase diagrams is presented in Figure 5.3. In this case, the vertical axis is once again the temperature, $T$, but the horizontal axes are now a quantity that has different values in the two coexisting phases – referred to as an *order parameter*. For the magnet, this quantity is the magnetization, $m$, which has a positive value in the up phase and a negative value in the down phase. This represents the ferromagnetic phase. For the liquid–gas system, the quantity that has different values in the liquid and gas phases is the density, $\rho$. In each case, note that a dome-shaped coexistence curve represents the coexistence region, and this region vanishes at the critical temperature. Above the critical temperature the magnetic system is in the paramagnetic phase, and the liquid–gas system is in the fluid phase.

Another way to understand the connection between magnets and fluids is to look at isotherms. In Figure 5.4 we show two different isotherms for a magnetic system. One is for a temperature well above the critical temperature in the *paramagnetic phase*. Here we see that the magnetization, $m$, of the

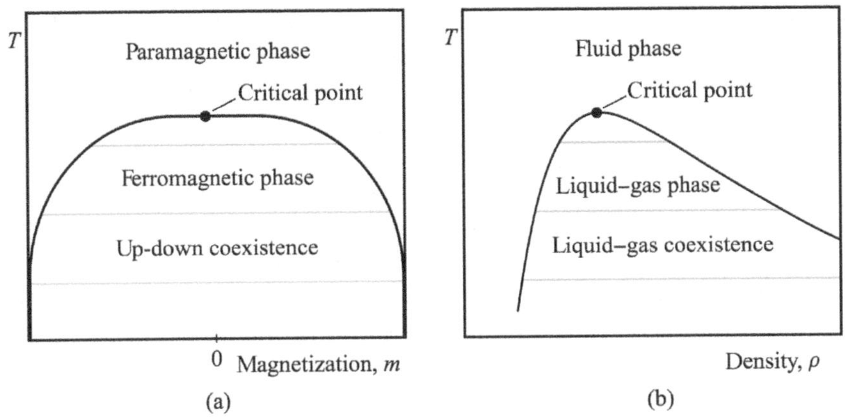


Figure 5.3 Coexistence phase diagrams for (a) a magnetic system and (b) a liquid–gas system. In the case of the magnet, the ferromagnetic phase is characterized by a difference between the up and down magnetizations. This difference vanishes at the critical point, and for higher temperatures the magnet is in the paramagnetic phase. In the liquid–gas system, the difference in density between the liquid and gas phases vanishes above the critical point, in the region referred to as the fluid phase.

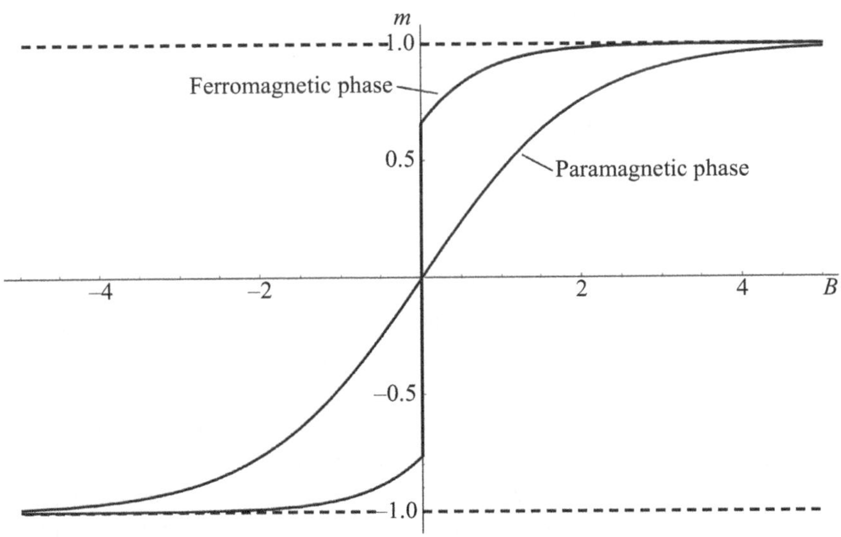


Figure 5.4 Two isotherms for a magnetic system. In the paramagnetic phase, the magnetization $m$ varies smoothly with the applied magnetic field $B$. In the ferromagnetic phase, the magnetization has coexisting positive and negative values of the magnetization at zero applied magnetic field.

magnet varies smoothly with the value of the magnetic field, $B$. In particular, in a near-zero magnetic field the magnetization varies linearly with the field and changes smoothly from positive to negative as the magnetic field changes sign. This is a hallmark of the paramagnetic phase.

The second isotherm is for a temperature below the critical point in the *ferromagnetic phase*. Here we see that as the magnetic field is reduced to zero, the magnetization decreases, but it does not go to zero – instead, it approaches a finite value. This residual magnetization, which occurs in the absence of an applied magnetic field, is referred to as the *spontaneous magnetization*. The fact that the magnetization remains finite, even when the applied magnetic field is zero, is why a bar magnet is often referred to as a permanent magnet.

In addition, notice that the spontaneous magnetization has a positive value when the applied magnetic field is reduced to zero from the positive magnetic field side, and a negative value when the field is reduced to zero from the negative side. As a result, it follows that at zero field there is a coexistence between up- and down-magnetized phases. This is a key characteristic of the ferromagnetic phase, and is analogous to the coexistence between the liquid and gas phases that is seen in low-temperature van der Waals isotherms.

We now turn to a bare-boned theoretical model that gives additional insight into magnetic systems. The model is simple enough to be studied in detail, but complex enough to reveal the basic features of real-world magnets.

## 5.2 The Ising Model for Magnetism

The Ising model (pronounced "I" "sing") is a theoretical model for magnetism that encompasses the key characteristics of magnetic behavior. It follows the maxim espoused by Albert Einstein that "Everything should be as simple as possible, but not simpler." The model was devised by Wilhelm Lenz in 1920, and was given to his graduate student, Ernst Ising, as the subject for his PhD dissertation. Ising published his results – an exact solution of the model on an infinite one-dimensional lattice – in 1925.

In keeping with the idea of simplicity, the Ising model represents a magnetic solid as a lattice of sites, labeled $i$, with each site occupied by an atom. The atoms in turn are represented by a "spin" variable, $s_i$, representing a free electron in the outermost shell. Since electrons are spin-1/2 particles, and can be either aligned or anti-aligned with a given direction, the Ising model assigns two values to the spin, corresponding to spin up and spin down. These values are defined to be plus one ($+1$) or minus one ($-1$). Thus, for each site $i$ on the lattice, there is a spin variable $s_i$ given by

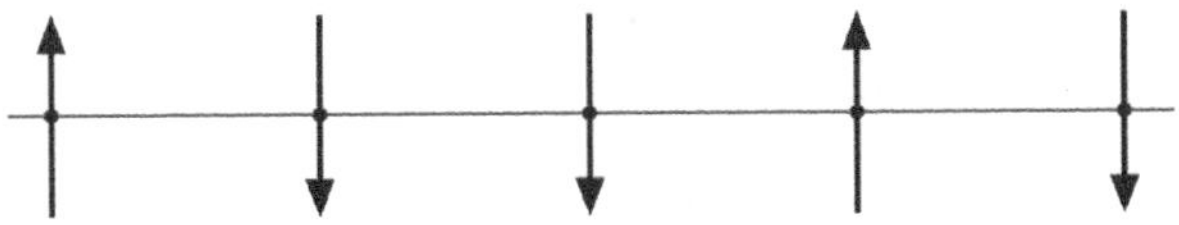

Figure 5.5 A one-dimensional representation of the Ising model. The sites $i$ on the lattice are indicated by "dots," and the Ising spin, $s_i$, at each site is indicated by an up arrow ($s_i = +1$) or a down arrow ($s_i = -1$).

$$s_i = +1 \quad \text{or} \quad s_i = -1.$$

An advantage of giving the Ising spins numerical values is that they can be easily manipulated in mathematical calculations.

Now, a magnetic *configuration* is defined as an assignment of +1 or –1 to each site of the lattice. An example for a one-dimensional system is given in Figure 5.5. If the majority of spins are –1, as in this figure, we say that the magnet has a negative magnetization.

The next step in defining the Ising model is to assign an energy to each spin configuration. First, the spins are imagined to be magnetic dipoles, and hence they have an energy when they are in an applied magnetic field $B$. The energy is

$$-\mu B s_i.$$

In this expression, $\mu$ is the magnetic moment of the spin. Notice that if the magnetic field $B$ is positive and $s_i = +1$, the energy of the system is less than if the magnetic field $B$ is positive and $s_i = -1$. Thus, when a spin is aligned with the magnetic field (that is, has the same sign) the result is a lowering of the energy.

The Ising model also incorporates interactions between neighboring atoms. When neighboring electrons are aligned in the same direction they must be far away from one another due to the exclusion principle, which states that one electron cannot be in the same state as another electron. As a result, their electrostatic energy of interaction is reduced. This interaction is referred to as the *exchange interaction*. We write the exchange interaction between neighboring spins, $s_i$ and $s_j$, as follows:

$$-J s_i s_j.$$

Notice that if $s_i = s_j$, and $J$ is positive, then the energy of the system is lowered. It follows that $J > 0$ corresponds to a ferromagnetic interaction; that is, it is an interaction that favors neighboring spins aligning with one another. Note that $J$ refers to an energy of interaction, and does not stand for joule, J, the unit of energy.

Combining these two energies, we can write the Hamiltonian $\mathcal{H}$ of an Ising system with the following expression:

$$\mathcal{H} = -J\sum_{<ij>} s_i s_j - \mu B\sum_i s_i.$$

The first term is the exchange interaction between neighboring spins, and the symbol $\sum_{<ij>}$ stands for "the sum over all nearest-neighbor pairs." In the second term, the symbol $\sum_i$ stands for "the sum over all sites." Thus, the second term is the total energy due to the applied magnetic field, $B$.

## The Boltzmann Factor and Probabilities

In analyzing a theoretical model, it's generally useful to work in terms of dimensionless quantities. For the Ising model, this takes the following form for the Hamiltonian:

$$-\beta\mathcal{H} = K\sum_{<ij>} s_i s_j + h\sum_i s_i. \quad (5.1)$$

In this expression, the term $\beta$ is the inverse of the thermal energy:

$$\beta = \frac{1}{k_B T}.$$

Note that $k_B$ is the Boltzmann constant,

$$k_B = 1.38 \times 10^{-23}\ \mathrm{J/K}.$$

The dimensionless factors in Eq. (5.1) are

$$K = \frac{J}{k_B T},$$

$$h = \frac{\mu B}{k_B T}.$$

We refer to $-\beta\mathcal{H}$ as the *reduced Hamiltonian*, and $K$ and $h$ as the *reduced interactions*.

The usefulness of the expression in Eq. (5.1), and in particular the inclusion of the minus sign in front of $\beta$, is apparent when we write what is known as the *Boltzmann factor*:

$$e^{-\beta\mathcal{H}}.$$

This quantity is key to statistical mechanics because the probability of a state with energy $E$ occurring is proportional to the Boltzmann factor. That is,

$$\text{probability} \propto e^{-\beta E} = e^{-E/k_B T}.$$

Thus, for example, if the energy $E$ is much greater than the thermal energy $k_\mathrm{B}T$, the probability of that state occurring is exponentially small. Thus, the advantage of defining the reduced Hamiltonian with a factor of $-\beta$ is that the result can be included directly into a Boltzmann factor.

As an example, suppose we have a system with just two states, with energies $E_1$ and $E_2$. The probability that the state with energy $E_1$ occurs is

$$P_1 = Ae^{-E_1/k_\mathrm{B}T}.$$

In this expression, $A$ is a proportionality constant, which will be determined shortly. Similarly, the probability for state 2 is

$$P_2 = Ae^{-E_2/k_\mathrm{B}T}.$$

Now, the total probability for the system occurring in one of these two states must be equal to 1; that is,

$$P_1 + P_2 = Ae^{-E_1/k_\mathrm{B}T} + Ae^{-E_2/k_\mathrm{B}T} = A\left(e^{-E_1/k_\mathrm{B}T} + e^{-E_2/k_\mathrm{B}T}\right) = 1.$$

From this it follows that the constant of proportionality is

$$A = \frac{1}{e^{-E_1/k_\mathrm{B}T} + e^{-E_2/k_\mathrm{B}T}} = \frac{1}{Z}.$$

In this expression, $Z$ is referred to as the *partition function.*

In general, for any system at all, the partition function is simply the sum of Boltzmann factors for all states of the system:

$$Z = \sum_{\text{all states}, n} e^{-E_n/k_\mathrm{B}T}. \tag{5.2}$$

With this result we can now write the probability of a given state as an equality. That is, the probability that a state with energy $E$ occurs is

$$\text{probability} = \frac{e^{-E/k_\mathrm{B}T}}{Z}.$$

This result will be used frequently in our calculations for the Ising model.

## Single-Spin Ising System

We start with the simplest possible Ising system – a single spin, $s$, in a magnetic field, $B$, and reduced magnetic field, $h$. Since there is only one spin in this system, we drop the subscript on $s$. The Hamiltonian, $\mathcal{H}$, and reduced Hamiltonian, $-\beta\mathcal{H}$, in this case are

$$\begin{aligned}\mathcal{H} &= -\mu B s. \\ -\beta\mathcal{H} &= hs.\end{aligned} \tag{5.3}$$

There is no nearest-neighbor interaction. This system – as simple as it is – will be useful in the next section when we consider the mean-field theory of magnetism.

In the single-spin system, there are just two states, spin up ($s = +1$) and spin down ($s = -1$). As a result, the partition function is

$$Z = \sum_{\text{all states}} e^{-\beta\mathcal{H}} = \sum_{\text{all states}} e^{hs} = e^{h} + e^{-h}. \tag{5.4}$$

We can use this result to calculate probabilities for the system. For example, consider the average value of the spin $s$, which we refer to as the *magnetization*, $m$. We denote this average with the symbol $\langle s \rangle$. That is,

$$m = \langle s \rangle.$$

If $m = +1$ the spin is always in the up state, and if $m = -1$ the spin is always in the down state. For intermediate values of $m$, between $-1$ and $+1$, the spin has a certain probability $P$ to be found up, and a corresponding probability $(1 - P)$ to be found down.

To calculate $m$, we start by writing out the relevant probabilities. The probability for $s = +1$ is

$$P = \text{probability for } (s = +1) = \frac{e^{h}}{Z}.$$

The probability for $s = -1$ is

$$\text{probability for } (s = -1) = \frac{e^{-h}}{Z} = 1 - P$$

Combining these results, the average value of $s$ is as follows:

$$m = \text{average value of } s = (+1)P + (-1)(1 - P).$$

This yields

$$m = \frac{e^{h} - e^{-h}}{Z}.$$

Applying our previous result for $Z$ in Eq. (5.4), we have

$$m = \frac{e^{h} - e^{-h}}{e^{h} + e^{-h}} = \tanh h.$$

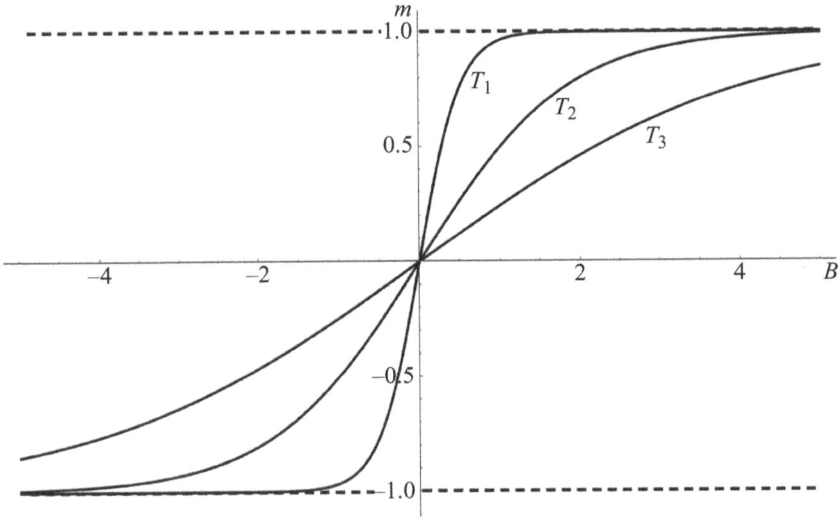


Figure 5.6 Single-spin magnetization, $m$, versus the magnetic field, $B$, for temperatures increasing from $T_1$ to $T_3$.

Notice that the maximum value for the magnetization is $m = +1$, as expected, and its minimum value is $m = -1$. When the reduced magnetic field is equal to zero, $h = 0$, the magnetization is also zero, $m = 0$.

In terms of the magnetic field $B$ and the temperature $T$ we can write $m$ as

$$m = \tanh h = \tanh \frac{\mu B}{k_B T}.$$

We plot $m$ for a fixed magnetic field $B$ and various temperatures in Figure 5.6. These are isotherms for the single-spin system, with a temperature that increases from $T_1$ to $T_3$.

Notice that all of the isotherms are smoothly varying with magnetic field, just as we saw earlier for the paramagnetic phase. There is never a coexistence between positive and negative magnetizations, as in a ferromagnet. Thus, there is no phase transition and no spontaneous magnetization for a single spin.

## Phase Transitions and Dimensionality

It's not surprising that a single spin has no phase transition, since a phase transition requires cooperative behavior between an immense number of interacting variables. For this reason, Ising studied not only the one-dimensional model, for simplicity, but the *infinite* one-dimensional model, in the hope that it would produce a phase transition. As it happens it doesn't – there is no phase transition in the one-dimensional Ising model. This was a major disappointment for Ising at the time of his dissertation.

To see why there is no phase transition in one dimension, consider the case of nearest-neighbor ferromagnetic couplings, $K > 0$, and zero magnetic field, $h = 0$. In Figure 5.7(a) we show the $T = 0$ ground state of an infinite one-dimensional lattice with these interactions. Notice that all the spins are aligned in the same direction. This means that all of the bonds between nearest neighbors are "satisfied"; that is, all are in their lowest-energy, aligned configuration.

As temperature is raised, the lowest-energy excitation above the ground state is a single "broken" bond, as shown in Figure 5.7(b) by the dashed connection between two spins. The spins to the left of this bond are all up, in their lowest-energy configuration, and the spins to the right are all down, also in their lowest-energy configuration. The two spins on either side of this broken (dashed) bond are anti-aligned, which yields a higher, unfavorable energy. This broken bond can occur anywhere on the lattice, however, resulting in a magnetization that vanishes because the system can have any fraction of spins up or down with the same energy. As a result, there is no finite-temperature state where the system has a finite magnetization, and hence there is no phase transition.

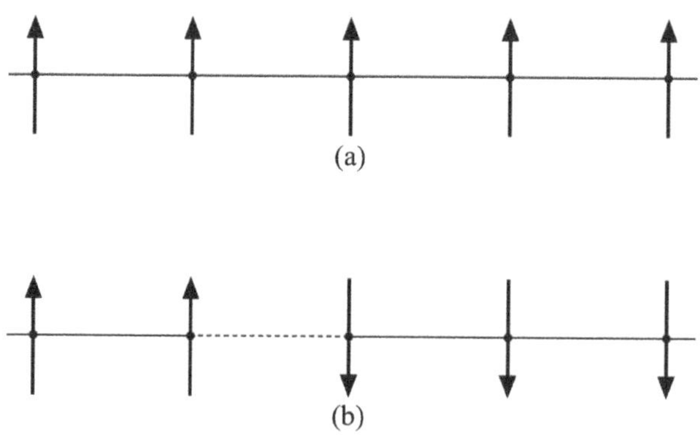


Figure 5.7 (a) The zero-temperature, ground state of the Ising model on a one-dimensional lattice. (b) The first excited state of the one-dimensional system as temperature is raised. There is one "broken" bond (dashed), which can occur anywhere on the lattice, resulting in zero magnetization.

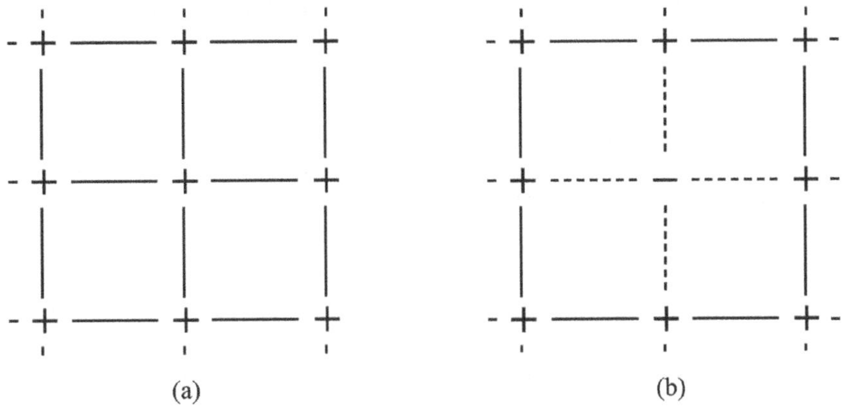


Figure 5.8 (a) The zero-temperature, ground state of the Ising model on a two-dimensional square lattice. (b) The first excited state of the two-dimensional system as the temperature is raised. There is one flipped spin and four "broken" bonds (dashed). The effect on the magnetization of the system is infinitesimal.

The situation in two dimensions is qualitatively different, as shown in Figure 5.8. In Figure 5.8(a) we see a two-dimensional, square lattice in its ground state, where all the spins are aligned in the plus direction. The lowest-energy excitation from the ground state is shown in Figure 5.8(b), where a single spin is flipped to the opposite direction. This results in four broken (dashed) bonds, but only a single spin in the opposite direction – as opposed to an infinite number of flipped spins in the corresponding one-dimensional case. The effect on the magnetization of the system is infinitesimal, and hence the system can maintain a finite magnetization even at a finite temperature, that is, the system can exhibit a phase transition.

An exact solution for the Ising model on the two-dimensional square lattice with zero magnetic field was derived by Lars Onsager in 1944, in what was truly a mathematical tour de force, and it does indeed show a finite-temperature phase transition, as expected. (For more details on the Ising model see *A Student's Guide to the Ising Model.*) So far, there is no exact solution for the Ising model on any three-dimensional lattice, but there are approximate solutions that make it clear that it too exhibits a phase transition. In Section 5.3, we present the mean-field approach to the Ising model, one of the most important approximation techniques for studying phase transitions in systems like the three-dimensional Ising model.

## 5.3 Mean-Field Theory for the Ising Model

To solve for the magnetic behavior of a three-dimensional Ising model as a model for physical bar magnets, we need to use approximation techniques, since no exact solutions exist. The first and simplest approximation method is due to Pierre Curie and Pierre Weiss. It was introduced in the early 1900s and is referred to as the Curie–Weiss theory, or more generally, as the *mean-field approximation*.

To understand this type of approximation, we start by imagining an Ising model in the ferromagnetic phase, that is, at a temperature below the critical temperature. If there is no applied magnetic field, there is still a spontaneous magnetization, $m$, in the system. This means that each spin on the lattice has a nonzero average, or mean, value. Specifically, for any site $i$ on the lattice, we have

$$\langle s_i \rangle = m \neq 0.$$

Again, this nonzero magnetization exists even though the applied magnetic field is zero.

Now, the basic idea of the mean-field approximation is that the *mean* magnetization of spins on the lattice produces an *effective magnetic field* that acts on other spins in the neighborhood. Specifically, consider a given spin on the lattice, $s_0$, as indicated in Figure 5.9. This spin is surrounded by nearest-neighbor spins, labeled $s_1$ to $s_4$, as shown on panel (a). Each of these spins interacts with $s_0$ with a strength $J$. In this figure we show the case of a two-dimensional square lattice for ease of illustration, but the same idea extends to

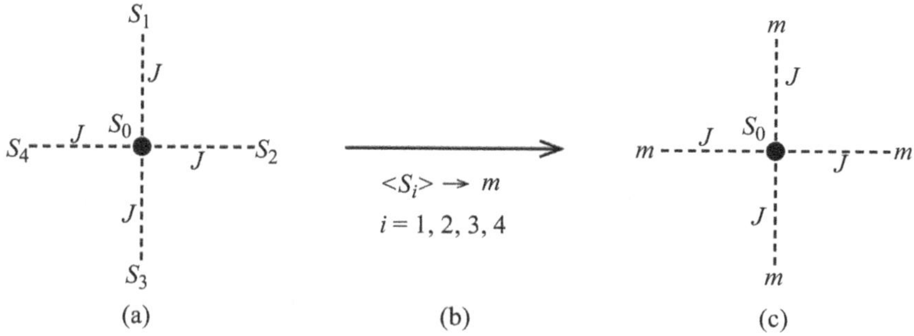


Figure 5.9 Graphical representation of the mean-field approximation for the Ising model. A single spin, $s_0$, interacts with couplings $J$ with its nearest neighbors, $s_1$ to $s_4$ in this case, which are replaced with the average value of a spin, $m$. The average value of $s_0$ is then calculated, and compared with $m$. When these values agree we have a self-consistent, mean-field solution.

three dimensions where, for a cubic lattice, there would be an additional nearest neighbor above and below the page.

To implement the mean-field idea, we replace each of the neighboring spins with its mean value $m$. This is illustrated Figure 5.9(c). These mean values interact with $s_0$, and with a strength $J > 0$ they tend to align $s_0$with the magnetization $m$.

Thus, the single spin $s_0$ behaves as if it were immersed in an effective magnetic field produced by its neighbors. Its Hamiltonian, $\mathcal{H}_0$, is given by

$$\mathcal{H}_0 = -qJms_0.$$

In this expression, $q$ is the number of nearest neighbors of $s_0$, four in this case, and is referred to as the *coordination number* of the lattice. Comparing with Eq. (5.3) for a single spin in a magnetic field, we see that the *effective* magnetic field experienced by $s_0$ is given by $\mu B_{\text{effective}} = qJm$, which is proportional to the magnetization $m$. The reduced effective Hamiltonian for $s_0$ is

$$-\beta\mathcal{H}_0 = h_{\text{effective}} s_0 = q\frac{J}{k_B T} m s_0.$$

Notice that we have now replaced the infinite lattice of spins with a single spin, $s_0$, which is representative of all the spins.

We can now express the basic idea of the mean-field approximation in the following way:

> The average value of $s_0$ – as produced by the effective magnetic field from the rest of the lattice – must be equal to the average value of the spins that produce the effective field. That is, we require that $\langle s_0 \rangle = m$. This is the *self-consistency condition* of mean-field theory.

We now implement this idea mathematically, using the basic principles of statistical mechanics.

To begin, we calculate the partition function for the single spin $s_0$. As in Section 5.2, this is done by summing over the two states of $s_0$, with the following result:

$$Z_0 = \sum_{\{s_0\}} e^{-\beta\mathcal{H}_0} = e^{q\frac{J}{k_B T}m} + e^{-q\frac{J}{k_B T}m}.$$

We can now write the probability $P$ that $s_0$ is $+1$ or $-1$ as follows:

$$P[s_0 = +1] = \frac{1}{Z_0} e^{q\frac{J}{k_B T}m(+1)} = \frac{1}{Z_0} e^{q\frac{J}{k_B T}m},$$

$$P[s_0 = -1] = \frac{1}{Z_0} e^{q\frac{J}{k_B T}m(-1)} = \frac{1}{Z_0} e^{-q\frac{J}{k_B T}m}.$$

Combining these results, the average value of $s_0$ is

$$\langle s_0 \rangle = \frac{1}{Z_0} \sum_{\{s_0\}} s_0 e^{-\beta \mathcal{H}_0} = \frac{(+1) e^{q \frac{J}{k_B T} m} + (-1) e^{-q \frac{J}{k_B T} m}}{e^{q \frac{J}{k_B T} m} + e^{-q \frac{J}{k_B T} m}}.$$

Simplifying yields

$$\langle s_0 \rangle = \frac{e^{q \frac{J}{k_B T} m} - e^{-q \frac{J}{k_B T} m}}{e^{q \frac{J}{k_B T} m} + e^{-q \frac{J}{k_B T} m}} = \tanh\left[q \frac{J}{k_B T} m\right].$$

Finally, applying the condition $\langle s_0 \rangle = m$ yields

$$m = \tanh\left[q \frac{J}{k_B T} m\right]. \tag{5.5}$$

This is the basic equation for this simple, single-spin version of mean-field theory.

## Properties of tanh(*x*) and Identification of the Critical Temperature

To get a better understanding of Eq. (5.5), we note some of the key properties of the hyperbolic tangent function, $\tanh(x)$, that appears on the right-hand side. First, $\tanh(x)$ is an odd function of its argument; that is,

$$\tanh(-x) = -\tanh(x).$$

As a result, $\tanh(0) = 0$. In addition, for large positive or negative values of the argument, $\tanh(x)$ approaches $+1$ and $-1$, respectively, as is appropriate for a function representing the magnetization, $m$. Finally, for small values of $x$, we can expand $\tanh(ax)$, where $a$ is an arbitrary constant, as

$$\tanh(ax) = ax + O(x^3).$$

In this expression, $O(x^3)$ means terms of order $x^3$ and higher. Thus, the slope of $\tanh(ax)$ near $x = 0$ is equal to $a$.

Each of these properties is illustrated in Figure 5.10, where we plot $\tanh(am)$ versus $m$ for $a = 0.5$ and 2.0. Also plotted for comparison is the left-hand side of Eq. (5.5), the straight line $y = m$. The intersections of the straight line and the tanh curves are solutions to Eq. (5.5).

Notice in Figure 5.10 that $y = m$ has a constant slope of 1. In comparison, the slope of $\tanh(0.5m)$ near the origin is $a = 0.5$, and the slope of $\tanh(2.0m)$ near the origin is $a = 2.0$. Clearly, then, for $a < 1$ there is only a single solution to Eq. (5.5), and it occurs at $m = 0$. For $a > 1$ there are three solutions: $m = 0$,

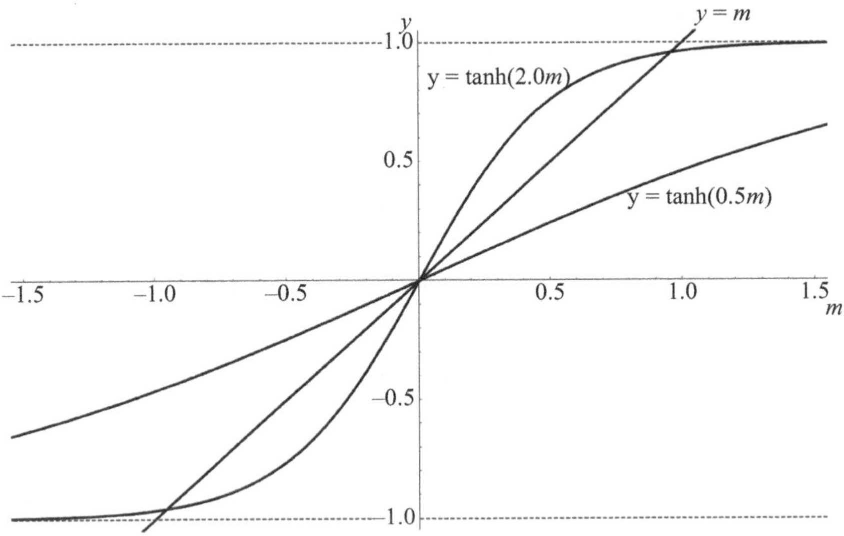


Figure 5.10 An illustration of basic properties of $y = \tanh(am)$ compared with the straight line $y = m$.

and $m = \pm$nonzero. The nonzero-$m$ solutions to Eq. (5.5) correspond to the spontaneous mean-field magnetization. In fact, as we shall see in Section 5.4, the nonzero solutions for $m$ give lower values of the free energy than the $m = 0$ solution, and hence are the equilibrium states of the system.

We can now apply these insights to Eq. (5.5). Expanding the right-hand side of the equation for small values of $m$ yields

$$\tanh\left[q\frac{J}{k_\mathrm{B}T}m\right] \approx q\frac{J}{k_\mathrm{B}T}m.$$

Thus, the initial slope of the right-hand side of Eq. (5.5) is $q\frac{J}{k_BT}$. If this quantity is greater than one there are two nonzero-$m$ solutions; if it is less than one the only solution is $m = 0$. It follows that we can associate an initial slope of 1 with the critical temperature, $T_\mathrm{c}$. In particular, we can write

$$q\frac{J}{k_\mathrm{B}T_\mathrm{c}} = 1.$$

Rearranging, the critical temperature can be expressed as

$$T_\mathrm{c} = q\frac{J}{k_\mathrm{B}}.$$

Notice that the critical temperature increases with the bonding strength, $J$, and with the number of nearest neighbors, $q$.

Using this result for the critical temperature, we can re-write Eq. (5.5) in a more convenient, dimensionless form. In particular, we start by inserting a $T_c$ in the numerator and denominator of the argument of the tanh function on the right-hand side of the equation. This yields

$$m = \tanh\left[q\frac{J}{k_B T}m\right] = \tanh\left[q\frac{J}{k_B T_c}\left(\frac{T_c}{T}\right)m\right].$$

Next, we note that $q\frac{J}{k_B T_c} = 1$, and hence we can write this as

$$m = \tanh\left[\left(\frac{T}{T_c}\right)^{-1} m\right]. \tag{5.6}$$

This is a *universal* result that applies to any system once we've divided the temperature by the critical temperature. We plot both sides of this equation in Figure 5.11, with the right-hand side plotted for three different values of $T/T_c$.

Figure 5.11 makes it clear that $T/T_c = 1$ represents the critical temperature. It is the dividing line between the high-temperature behavior ($T/T_c \geq 1$), where the only solution (intersection between the curve and the straight line) is $m = 0$, and the low-temperature behavior ($T/T_c < 1$), which has two nonzero-$m$ solutions symmetrically placed on either side of $m = 0$.

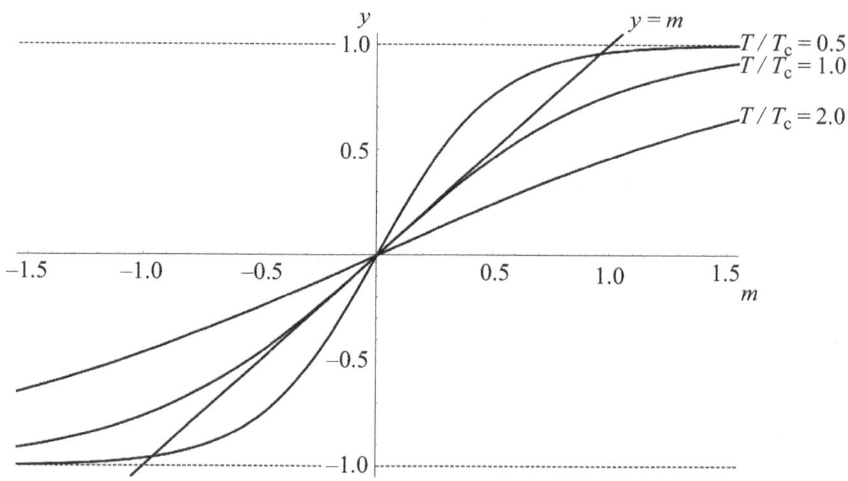


Figure 5.11 A plot of the left- and right-hand sides of Eq. (5.6). For $T/T_c \geq 1$ there is a single solution at $m = 0$; for $T/T_c < 1$ there are two additional solutions for nonzero $m$.

### Cobwebbing and the Relaxation Method

Our task now is to find the nonzero solutions to Eq. (5.6). Unfortunately, there is no analytic solution to this equation. Fortunately, this isn't a problem because, in a sense, the equation solves itself.

What do we mean by this? Well, we can see this by considering Eq. (5.6) as an iterative process. For example, suppose we start with an initial magnetization $m_0$. Now, let's substitute $m_0$ into the right-hand side of Eq. (5.6), and call the result on the left-hand side of the equation $m_1$, the first iteration of the magnetization. That is,

$$m_1 = \tanh\left[\left(\frac{T}{T_c}\right)^{-1} m_0\right].$$

Repeating this process yields the second iteration of the magnetization, $m_2$:

$$m_2 = \tanh\left[\left(\frac{T}{T_c}\right)^{-1} m_1\right].$$

In general, we have

$$m_{n+1} = \tanh\left[\left(\frac{T}{T_c}\right)^{-1} m_n\right].$$

It turns out that continuing with these iterations leads to a *fixed point* $m^*$ defined as follows:

$$m^* = \tanh\left[\left(\frac{T}{T_c}\right)^{-1} m^*\right].$$

Clearly, the fixed point $m^*$ is the desired nonzero solution to Eq. (5.6).

To see how this works in practice, let's consider a specific example. We'll start with $m_0 = 0.4$ and $T/T_c = 0.5$, as shown in Figure 5.12. The first iteration yields

$$m_1 = \tanh\left[\left(\frac{T}{T_c}\right)^{-1} m_0\right] = \tanh\left[(0.5)^{-1}(0.4)\right] = 0.665037\ldots$$

We also show $m_1$ in Figure 5.12. The next iteration yields

$$m_2 = \tanh\left[\left(\frac{T}{T_c}\right)^{-1} m_1\right] = \tanh\left[(0.5)^{-1}(0.665037\ldots)\right] = 0.868778\ldots$$

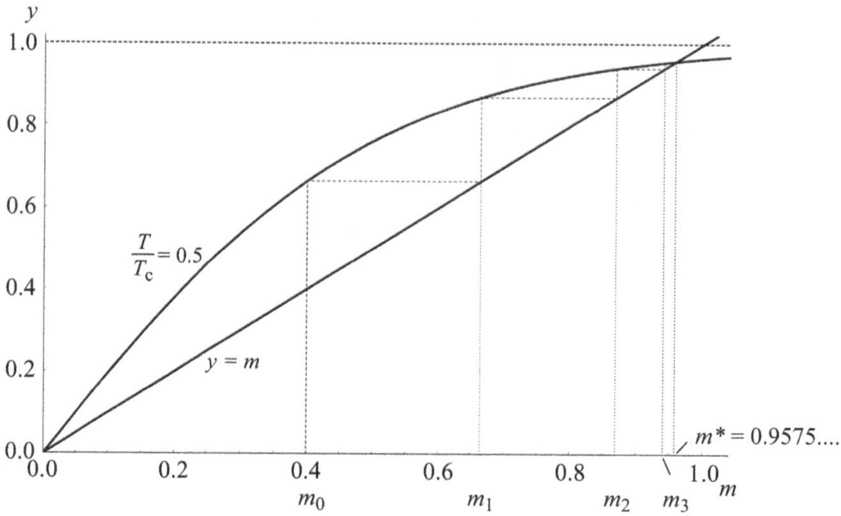


Figure 5.12 Cobwebbing from an initial value, $m_0 = 0.4$, to the final value, $m^* = 0.9575\ldots$ for $T/T_c = 0.5$. Any finite value for $m_0$ will *relax* to the same value of $m^*$ for $T/T_c = 0.5$.

Taking one more step in the process we have

$$m_3 = \tanh\left[\left(\frac{T}{T_c}\right)^{-1} m_2\right] = \tanh\left[(0.5)^{-1}(0.868778\ldots)\right] = 0.939942\ldots$$

As we can see in Figure 5.12, the iterations of the magnetization are quickly approaching the fixed point given by

$$m^* = \tanh\left[\left(\frac{T}{T_c}\right)^{-1} m^*\right] = \tanh\left[(0.5)^{-1}(0.957504\ldots)\right] = 0.957504\ldots$$

This is the solution to Eq. (5.6) for $T/T_c = 0.5$, and all we have to do is keep substituting iterations of the magnetization back into the tanh function until $m^*$ is reached to any desired level of accuracy. The equation solves itself.

We refer to the graphical representation of the iterative process shown in Figure 5.12 as *cobwebbing*. This is because the dashed lines connecting one iteration to the next look like a web spun by a small spider going back and forth between the tanh curve and the straight line. Any reasonable choice for $m_0$ will cobweb itself to the same fixed point for this value of $T/T_c$.

This method of cobwebbing is often referred to as the *relaxation method* for solving an equation. An initial value for $m_0$ is chosen, and the resultant value of $m_1$ is calculated. With the temperature as low as $T/T_c = 0.5$, the value of $m_1$ is greater than $m_0$ because the interactions with nearest neighbors are strong enough to produce a large magnetization. So the original choice of $m_0 = 0.4$ was too low, and the iterations cause the system to *relax back* to the larger value that is consistent from one iteration to another. The same type of relaxation from initial guess to final solution occurs when calculating the electric potential by numerically solving Laplace's equation.

## Numerical Solution for the Magnetization

We've seen that the fixed-point magnetization for $T/T_c = 0.5$ is $m^* = 0.9575\ldots$ Carrying out the same relaxation calculation for other values of $T/T_c$ gives the values presented in Table 5.1. Notice that for low values of $T/T_c$ the magnetization remains rather large, that is, close to 1, but the magnetization drops off quickly once $T/T_c$ reaches values of about 0.7.

The values given in Table 5.1 are plotted with solid dots in Figure 5.13. The smooth curve connecting the dots is the complete solution for all values of $T/T_c$.

In Figure 5.14 we plot the magnetization again, this time with the temperature on the vertical axis and the magnetization on the horizontal axis. We also show both the positive and negative values for the magnetization. Note that the dome-shaped coexistence curve for the magnetization vanishes at the critical point, as expected. We next turn to the detailed behavior of the magnetization in the vicinity of the critical point.

Table 5.1 *Mean-field solutions for a variety of values of $T/T_c$*

| $T/T_c$ | $m$ |
|---|---|
| 0 | 1 |
| 0.1 | 1.0000 |
| 0.2 | 0.9999 |
| 0.3 | 0.9974 |
| 0.4 | 0.9856 |
| 0.5 | 0.9575 |
| 0.6 | 0.9073 |
| 0.7 | 0.8286 |
| 0.8 | 0.7104 |
| 0.9 | 0.5254 |
| 1 | 0 |

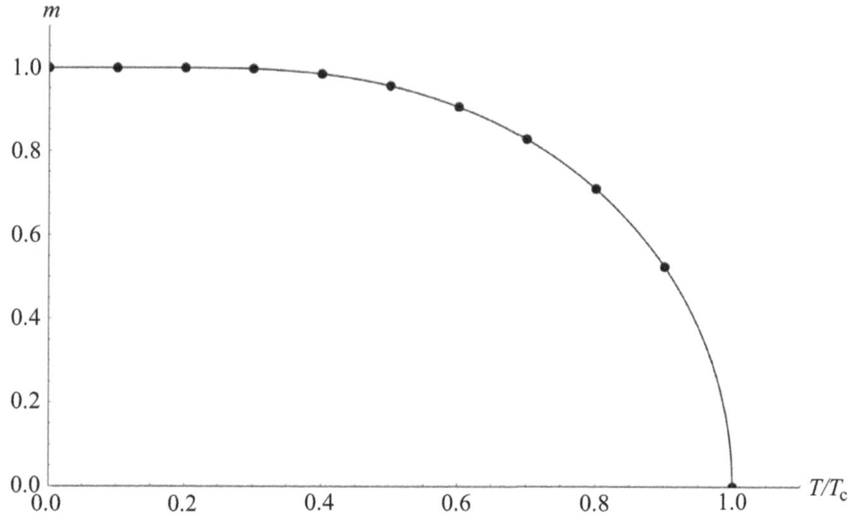


Figure 5.13 Self-consistent solutions from Table 5.1 (dots) and complete solution (solid curve) for the mean-field magnetization for the Ising model.

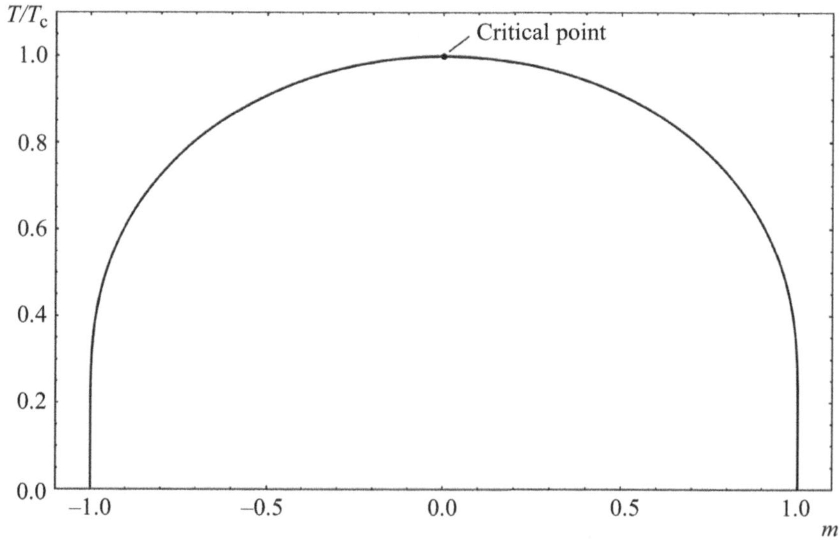


Figure 5.14 The mean-field temperature-versus-magnetization phase diagram for the Ising model. This phase diagram corresponds with that shown in Figure 5.3(a). At the critical point the dome-shaped coexistence region vanishes. Above that temperature the system is paramagnetic; below it the system is ferromagnetic.

## Critical Exponent for the Magnetization and Order Parameters

The behavior of the coexistence curve near the critical temperature turns out to be a particularly significant aspect of critical behavior. One might wonder, for instance, how exactly the magnetization vanishes as the critical temperature is approached? Is it as a power law dependence, an exponential dependence, or does it take some other functional form? We address this issue now, and, as we shall see, the answer has wide-ranging implications.

To start, we expand the tanh function to one higher power than we did earlier in this section. This yields

$$\tanh(ax) = ax - \frac{1}{3}(ax)^3 + O(x^5).$$

Applying this to Eq. (5.6) we find

$$m = \tanh\left[\left(\frac{T}{T_c}\right)^{-1} m\right] \cong \left(\frac{T}{T_c}\right)^{-1} m - \frac{1}{3}\left(\frac{T}{T_c}\right)^{-3} m^3.$$

Canceling one power of $m$ and rearranging gives

$$m^2 = 3\left(\frac{T}{T_c}\right)^3\left(\left(\frac{T}{T_c}\right)^{-1} - 1\right) = 3\left(\frac{T}{T_c}\right)^2\left(1 - \left(\frac{T}{T_c}\right)\right).$$

This result applies to any system once we reduce the temperature by the critical temperature.

Now, as a convenient measure of the "distance" from the critical temperature, we introduce the reduced temperature $t$ defined as follows:

$$t = 1 - \left(\frac{T}{T_c}\right).$$

We see that $t = 0$ defines the critical point, and $t$ is positive for temperatures below the critical temperature. In terms of $t$, we can write the magnetization near the critical point as

$$m^2 = 3(1-t)^2 t.$$

Solving for the magnetization $m$ yields

$$m = \sqrt{3}(1-t)t^{1/2} \underset{t \to 0}{\rightarrow} \sqrt{3}t^{1/2}.$$

Thus, the dominant dependence of $m$ on reduced temperature $t$ is

$$m \approx t^{1/2}.$$

So the magnetization vanishes as a power law, and the power is ½. By convention, this power – referred to as a *critical exponent* – is defined to be $\beta$. That is,

$$m \approx t^{\beta}.$$

Thus, for this single-spin mean-field solution, the critical exponent for the magnetization is

$$\beta = \frac{1}{2}.$$

Other critical exponents describe the dependence of other quantities near the critical point.

Now, as it turns out, this result applies to much more than just the single-spin calculation presented here. In fact, *any* mean-field approximation, no matter how involved and complex, yields exactly the same value for $\beta$.

Not only that, but we can generalize this result by noting that $m$ is a measure of "how far" the system is from criticality. As mentioned earlier in this chapter, we refer to $m$ as the *order parameter* of the system. In general, an order parameter is zero in the high-temperature phase of a system, whether it is a magnet, a liquid–gas, or some other system, and becomes nonzero when the temperature is lowered below the critical temperature. It turns out that the critical exponent $\beta$ for the order parameter is the same in mean-field theory for any system that undergoes a critical phase transition.

For example, in the van der Waals theory of the liquid–gas phase transition, the order parameter is the *difference in density* between the liquid and gas phases. This density difference is zero for temperatures above the critical point but is finite for temperatures below it. If we look at how the difference goes to zero as the critical point is approached, we again find that it vanishes as a power law, and with the same critical exponent, $\beta = 1/2$. This is an example of what is referred to as the *universality of critical behavior*, where systems that are apparently very different show the same behavior near the critical point.

Mean-field theory shows that many different systems have the same critical exponents but, unfortunately, the critical exponents predicted by mean-field theory do not agree with experiment. The reason is that mean-field theories ignore fluctuations, and instead focus on the average (mean) behavior of a system. It turns out, as seen in renormalization-group theory, that fluctuations are crucial in determining critical behavior, and in yielding universality; in fact, fluctuations can become unbounded near critical points. For example, the critical exponent $\beta$ is ½ in mean-field theory, but is measured experimentally to be closer to 0.326. Renormalization-group calculations give exponents much closer to those observed experimentally. The insights gained from fluctuation theories of phase transitions can be explored in more detail in *A Student's Guide to the Ising Model.*

## 5.4 Landau Expansions

Now that we've seen how the order parameter of a system characterizes a phase transition, we turn to a phenomenological approach due to Lev Landau. In this approach, we expand the free energy in a power series for small values of the order parameter. Analyzing such a Landau expansion gives additional insight into the mechanics of phase transitions.

### Zero Magnetic Field

To begin, consider an Ising system with zero magnetic field. In this case, the system is up–down symmetric, that is, flipping all of the spins in the system results in no change of its properties. It follows that an expansion of the free energy of the system in powers of $m$ must contain only even powers, so that changing $m$ to $-m$ has no effect. Thus, to the lowest relevant power, we have

$$f = am^2 + bm^4. \tag{5.7}$$

In this expansion the parameter $a$ can be positive or negative, and in all the cases we consider the parameter $b$ is positive, which ensures that the free energy has a finite minimum value. Higher powers in the expansion are not needed to obtain the basic results of this approach.

We plot $f$ in Figure 5.15 for a variety of values of $a$. In panel (a) we show the case for $a > 0$, which results in a single minimum in the free energy at $m = 0$. This represents the disordered (zero order parameter) phase of the system. In panel (b) we show the case for $a = 0$, which gives a very flat minimum at $m = 0$. Reducing this parameter to negative values, $a < 0$, as shown Figure 5.15(c), we see a qualitative difference in the free energy – now there are two symmetric minima at finite values of $m$. These represent the ordered (finite order parameter) phase of the system. This agrees with the discussion regarding the mean-field solutions in Figure 5.11, where we pointed out that the finite-$m$ solutions correspond to the minima in the free energy and represent the equilibrium states of the system.

We note that $a = 0$ represents the critical point of the system. After all, it is the dividing point between the region ($a \geq 0$), with only $m = 0$ solutions, and the region ($a < 0$) with finite-$m$ solutions. The fact that the free energy is so

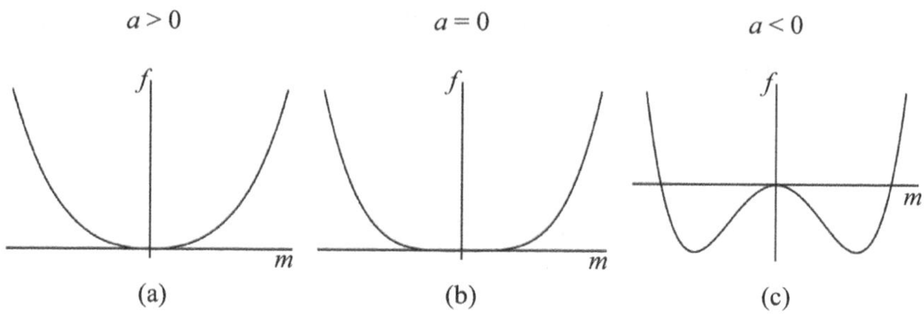


Figure 5.15 Plots of the expansion for the free energy $f$ for various values of $a$, and for a positive value of $b$. (a) We see that $a > 0$ corresponds to high temperature, with a single minimum at $m = 0$. (b) We associate $a = 0$ with the critical point. (c) For $a < 0$ the lowest values of the free energy are at finite values of $m$, indicating the ordered phase.

flat at the critical point indicates that the system is slow to move back to $m = 0$ if it is moved slightly away from equilibrium. This is referred to as *critical slowing down.*

## Critical Exponent

Using the Landau expansion, we can examine in detail the way the order parameter goes to zero as the critical temperature is approached. Specifically, near the critical point we note that $a$ is changing sign. Thus, we can write for $a$ near the critical temperature, assuming it has a linear dependence, the following expression:

$$a = -a't.$$

In this expression the quantity $a'$ is positive, so that for temperatures above the critical temperature ($t < 0$) we have $a = -a't > 0$, and for temperatures below the critical temperature ($t > 0$) we have $a = -a't < 0$.

Now, to find the finite-$m$ solutions below the critical temperature, that is, to find the values of $m$ that give the global minima of the reduced free energy and to see how these values of $m$ go to zero as $t$ goes to zero, we set the derivative of $f$ with respect to $m$ equal to zero. This yields

$$\frac{df}{dm} = 2am + 4bm^3 = 0.$$

Clearly, there is always an $m = 0$ solution to this equation. For the finite-$m$ solutions we have

$$m = \sqrt{\frac{-2a}{4b}} = \sqrt{\frac{2a't}{4b}}.$$

Thus, we see that $m$ vanishes as the square root of the reduced temperature $t$. That is,

$$m \sim t^{1/2}.$$

It follows that the critical exponent for the order parameter is

$$\beta = \frac{1}{2}.$$

This agrees with the result obtained in other mean-field theories.

## Finite-Field Behavior

Next, we add the effects of an external magnetic field. Recalling that the energy associated with the reduced magnetic field $h$ is given by $-hm$, we have

$$f = -hm + am^2 + bm^4.$$

Note that the magnetic field term breaks the up–down symmetry of the system. This is evident in Figures 5.16 and 5.17. We can also add an arbitrary constant to $f$, which simply adjusts the zero level of energy.

First, in Figure 5.16, we consider the case of temperatures greater than the critical temperature; that is, with $a > 0$. When $h < 0$ we see that the minimum of the reduced free energy is at a negative value of $m$. As we change $h$ to zero, and then to positive values, we see that the minimum of $m$ also goes to zero and then to positive values. Thus, the equilibrium state of the system has a value of $m$ that varies smoothly with the value of the reduced magnetic field.

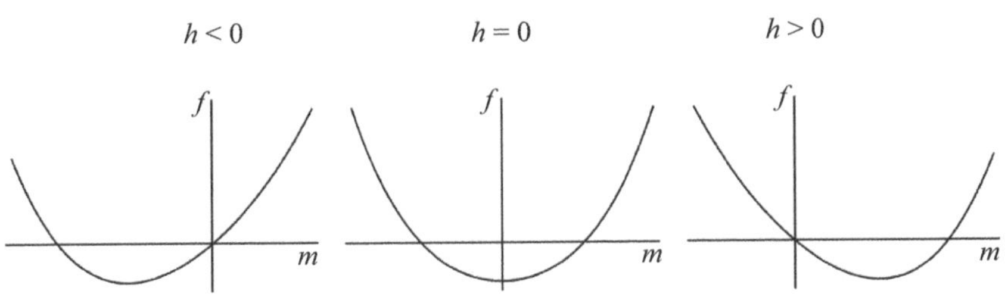


Figure 5.16 The reduced free energy $f = -1 - h\, m + m^2 + m^4$ for a variety of values of the reduced magnetic field $h$ and a temperature greater than the critical temperature. We see that the sign of $m$ at the minimum varies smoothly with the sign of the reduced magnetic field, as expected for the paramagnetic phase.

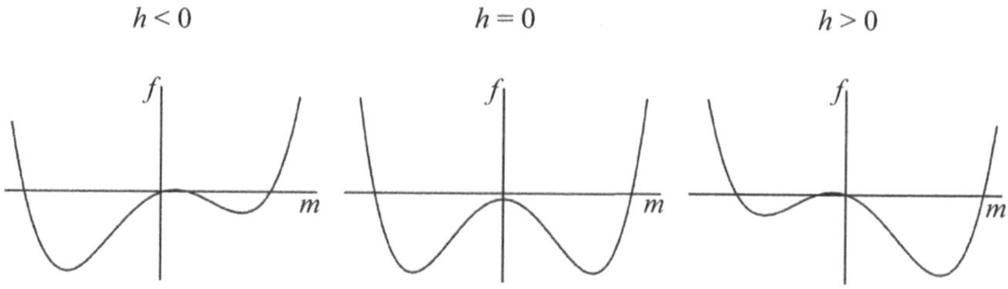


Figure 5.17 The reduced free energy $f = -0.2 - h\,m - 2m^2 + m^4$ for a variety of values of the reduced magnetic field $h$ and a temperature less than the critical temperature. We see that at $h = 0$ the finite-$m$ solutions are $m = \pm$finite, as expected for the ferromagnetic phase.

This is precisely what is expected for a system in the high-temperature paramagnetic phase.

In Figure 5.17, we show the reduced free energy $f$ for a temperature below the critical temperature; that is, with $a < 0$. For $h > 0$ and $h < 0$ the global minimum of the free energy occurs at a positive and negative value of $m$, respectively. At $h = 0$ the minima with positive and negative values of $m$ have the same value of $f$, and hence they coexist, as expected for a ferromagnetic system.

Note that for finite fields we have both global minima and local minima. The global minima correspond to the equilibrium states of the system, but the local minima can have interesting consequences as well. We explore this next.

## Hysteresis

Imagine a system corresponding to the global minimum of the left-most plot in Figure 5.17. We show this state, which has a negative value of $h$ and $m$, with a black dot in the lower-left corner of Figure 5.18. As $h$ is increased to zero, this minimum rises and becomes equal to the minimum corresponding to a positive value of $m$. We see this in the lower-center plot in Figure 5.18.

These two states are now in coexistence, but the system may stay in the left-most minimum even as $h$ is increased to positive values. It's like holding a ball in your cupped hand at the same level as the floor, and then raising it to heights above the floor. Similarly, as $h$ is increased, or your hand is raised, the system, and the ball, can remain "trapped" in a local minimum, as shown in the lower-right plot in Figure 5.18.

The system can remain there, in a metastable state (like in Figure 4.6) unless it is "jostled," like something bumping your hand, in which case the system (or ball) can fall to the global minimum. This is shown in the upper-right plot in Figure 5.18. Here we see that $m$ has increased to a positive value, and the

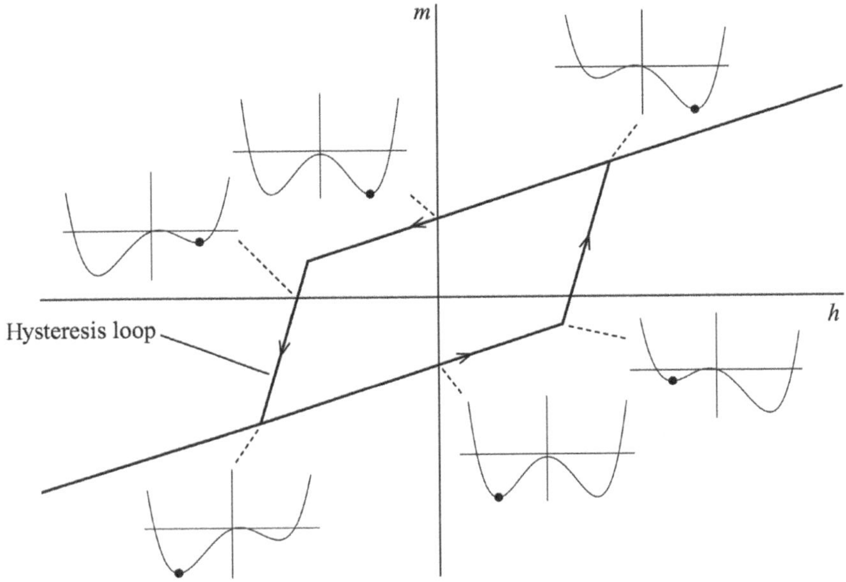


Figure 5.18 Hysteresis loop in magnetism. Each small plot shows the reduced free energy, $f$, plotted versus the reduced magnetization, $m$. As $h = 0$ is crossed in the positive direction, the state of the system (the black dot) is "trapped" in a local minimum above the global minimum. Eventually the system drops to the global minimum, as shown in the upper-right plot. This same process occurs again on the way back down in values of $h$, resulting in a hysteresis loop.

system is in its equilibrium state. If we now lower the reduced magnetic field past zero the system can once again become "trapped" in a metastable state, this time at a positive value of $m$, as shown in the upper-left plot in Figure 5.18. Another jostle can send it tumbling to the global minimum, and back to its starting point with a negative value of $m$, as in the lower-left plot.

Notice that we have just completed a closed circuit in the $m$-versus-$h$ plot. This closed circuit is referred to as a *hysteresis loop*. In general, the effect of having different results in a system depending on its past *history* is referred to as *hysteresis*.

The effect of hysteresis is probably most common in magnetic systems in the ferromagnetic phase where there is a discontinuous phase transition, but it can also be observed in other kinds of systems. In fact, a hallmark of discontinuous phase transitions in general, like the liquid–vapor phase transition, is the possibility of hysteresis effects.

Hysteresis can also be seen in common real-world situations. We show an example related to motor boating in Figure 5.19.

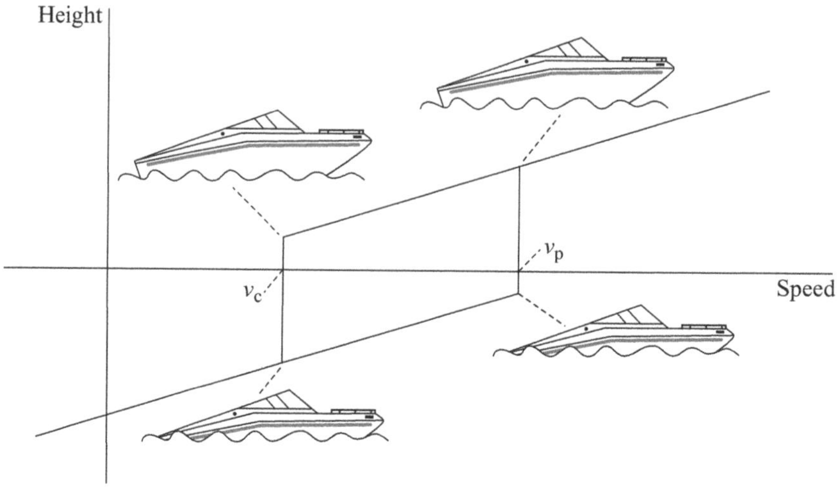


Figure 5.19 Hysteresis in motor boating. The boat has to reach a higher speed, $v_p$, to rise up on a plane than the speed where it drops back down to cruising level, $v_c$, as it slows down. This results in a hysteresis loop.

When you pilot a powerboat, you start with a small speed, where the boat is low in the water and traveling in a mode referred to as "cruising," indicated with the subscript c. This is shown in the lower images in Figure 5.19. As the speed of the boat is increased, you eventually reach a speed, $v_p$, where the boat rises up on the water and skims at high speed over its surface. This mode of travel is referred to as "planing," and is indicated with the subscript p.

What makes this transition between modes particularly interesting is that once the boat is on a plane, it stays there at lower speeds than $v_p$ when slowing down, as shown in the upper images in Figure 5.19. At the lower speed $v_c$ the boat drops back down to the cruising mode, as shown in the lower-left image in the figure. The result is a hysteresis loop, and a situation where the mode of travel of a boat is dependent on its past history – the powerboat shows hysteresis.

## 5.5 Problems

5.1 **Mean-field solution.** Consider the single-spin, Ising model mean-field solution given in Eq. (5.6).
**(a)** Find $m_1$ for $T/T_c = 0.55$ and $m_0 = 0.26$.
**(b)** Is $m^*$ for $T/T_c = 0.55$ greater than, less than, or equal to $m^* = 0.9575$ for $T/T_c = 0.4$? Explain your reasoning.
**(c)** Find $m^*$ for $T/T_c = 0.55$.

5.2 **Mean-field solution.** Consider the single-spin, Ising model mean-field solution given in Eq. (5.6).
**(a)** Find $m_1$ for $T/T_c = 0.50$ and $m_0 = 0.33$.
**(b)** Is $m^*$ for $T/T_c = 0.50$ and $m_0 = 0.33$ greater than, less than, or equal to $m^*$ for $T/T_c = 0.50$ and $m_0 = 0.26$? Explain your reasoning.
**(c)** Find $m^*$ for $T/T_c = 0.50$.

5.3 **Landau expansion.** Consider the Landau expansion given in Eq. (5.7) with $a = 1$ and $b = 1$.
**(a)** How many minima does this function have?
**(b)** What is the location and value of $f$ at each of these minima?

5.4 **Landau expansion.** Consider the Landau expansion given in Eq. (5.7) with $a = -1$ and $b = 1$.
**(a)** How many minima does this function have?
**(b)** What is the location and value of $f$ at each of these minima?

5.5 **Finite field.** The single-spin result in Eq. (5.6) was obtained for the case of zero external magnetic field. Show that in the presence of an external reduced magnetic field $h$ the self-consistent equation is

$$m = \tanh\left[h + \left(\frac{T}{T_c}\right)^{-1} m\right].$$

5.6 **Finite field.** Referring to the result given in Problem 5.5, let $T/T_c = 0.75$. Find **(a)** $m^*$ with $h = 0$ and **(b)** $m^*$ with $h = 0.25$.

# 6
# Binary Mixtures

In previous chapters, we considered phase transitions involving a single substance such as water, carbon dioxide, or iron. Now we turn to cases where the system of interest consists of two different substances, that is, a binary mixture. The mixture can be in the vapor, liquid, or solid phase, or it can exhibit coexistence between any two of these phases.

In what follows we start at high temperature, in the vapor phase, and work our way down to the liquid and solid phases. Finally, at the end of the chapter, we revisit the liquid phase to investigate a bizarre situation where the high-temperature phase appears to be more ordered than the low-temperature phase. We refer to this as an "upside-down" phase transition.

## 6.1 Liquid–Vapor Phase Transitions

We start with phase transitions between the vapor and liquid phases of a binary mixture. At high temperature the mixture is a vapor, with the two substances uniformly intermingled. As the temperature is lowered the mixture condenses to a liquid. In the cases considered in this section, the liquid is also a uniform mixture of the two components, just like the vapor. This is referred to as a *miscible* (mixed) liquid mixture. In the last two sections of the chapter we extend our considerations to liquid mixtures that do not mix uniformly; that is, to *immiscible* liquids.

### Lens-Shaped Phase Diagrams

Consider a mixture consisting of two different substances, which we refer to as A and B. Figure 6.1 shows a specific case in a temperature-versus-composition phase diagram. In this case, component A represents cyclohexane and

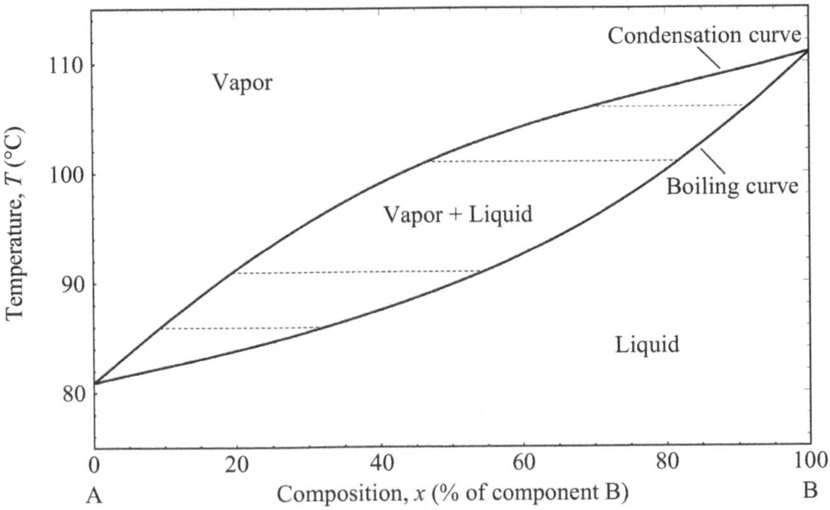


Figure 6.1 Temperature-versus-composition phase diagram for cyclohexane (A) and toluene (B). The composition, on the horizontal axis, is simply the percentage of component B in the mixture. At high temperature the system is in the vapor phase, and at low temperature it is in the liquid phase. In between, in the lens-shaped region, the two phases coexist. The upper curve of the lens-shaped region is where the vapor condenses; the lower curve is where the liquid boils.

component B represents toluene. The composition $x$ on the horizontal axis is the percentage of B in the mixture. Notice that in the upper portion of the phase diagram the system is in the vapor phase; in the lower portion it is in the liquid phase. In between there is a *lens-shaped region* representing coexistence between these two phases.

Let's take a closer look at the phase diagram in Figure 6.1. First, if the mixture is pure A, as on the left-hand side of the phase diagram, we see that the liquid boils to a vapor at a temperature of about 81 °C. On the other hand, if the system is pure B, as on the right-hand side of the phase diagram, the liquid–vapor phase transition occurs at about 111 °C. Thus, component A is said to be more *volatile* than B – that is, it boils away more readily and at a lower temperature. It follows that the B molecules attract other molecules with a greater strength than the A molecules do.

Now, if we start at pure A and add a small percentage of B to the system, the fact that the B molecules have stronger interactions means that the phase transition from liquid to vapor occurs at a higher temperature. In addition, the more volatile, weakly interacting A molecules are more likely to be found in the low-density vapor phase, and the strongly interacting B molecules gather together preferentially in the denser liquid phase. The result is a coexistence of a lower composition (A-rich) vapor phase with a higher composition (B-rich) liquid phase. This is indicated by the dashed, horizontal lines crossing the lens-shaped region.

We refer to the lower curve bounding the lens-shaped region as the *boiling curve*. It gives the temperature, for any given composition, where the liquid mixture begins to boil and give off vapor as it is heated. Similarly, the upper curve on the lens-shaped region indicates the temperature where the vapor phase begins to condense to form a liquid as the mixture is cooled. We refer to it as the *condensation curve*.

## Closed Systems

Imagine a *closed container* holding a mixture of A and B molecules. It follows that the composition $x$ of this mixture must remain constant, regardless of the temperature. In Figure 6.2 we show a specific case, again for cyclohexane (A) and toluene (B). The vertical dashed line indicates a particular system that is prepared with a composition of 30% A molecules and 70% B molecules.

Let's start at point 1 in Figure 6.2, which is at a temperature of $T = 90\,°C$. At this point the system is a uniform liquid. When we raise the temperature of the liquid, it begins to boil at point 2, at a temperature of 96 °C. The vapor that

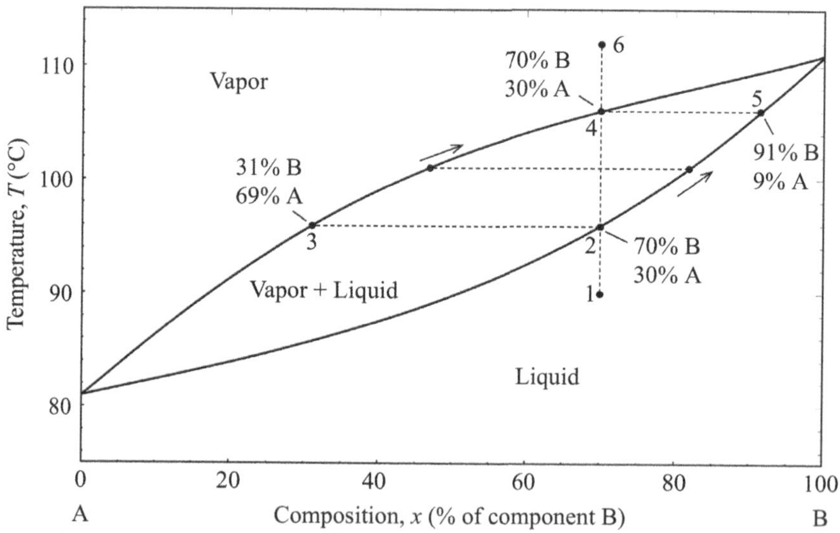


Figure 6.2 The behavior of a closed binary system, where the vertical line represents the particular system comprising 30% A molecules and 70% B molecules.

boils off from the liquid has the much lower composition of 31% B, as indicated by point 3.

As the temperature is raised further, the composition of both the liquid and the vapor rise, as indicated by the arrows on either side of the lens-shaped region. Finally, at point 4, the last bit of liquid boils away to a vapor, which now has a composition of 70% B. The last drop of liquid has a composition of 91% B, as indicated by point 5. If we continue to raise the temperature we reach point 6, where the system is now a pure vapor, with a composition of $x = 70\%$ B.

The key point to notice in this phase diagram is the discontinuity in composition as we traverse the lens-shaped region. This applies whether we cross the lens-shaped region from the bottom upward, resulting in the boiling of a liquid, or from the top to the bottom, which results in a vapor condensing to a liquid.

### Distillation and Fractional Distillation

In Figure 6.2, we see that when the A-B liquid boils in a closed system, as at point 2, the vapor that comes off has a much higher concentration of A molecules than did the original liquid. In particular, the original liquid in this case has a concentration of 30% A molecules, and the vapor has the much higher concentration of 69% A. This is a simple example of *distillation*, where boiling a liquid mixture results in one component being enriched in the vapor.

To see the process of distillation more clearly, consider an *open system* – where the vapor boils freely from the liquid and is not contained in an enclosure. We show this for the same A-B mixture from Figure 6.2 in Figure 6.3. Once again, we imagine heating a mixture with 30% A to the boiling temperature at 96 °C. As mentioned, the vapor boiling off has a concentration of 69% A.

This is a significant enrichment of the A concentration, but we can do even better. If the vapor with 69% A is captured, and then brought to its boiling point, which is roughly 86 °C, we find a new vapor boiling off with an even higher concentration of 90.7% A, as shown in Figure 6.3. This process can be repeated, giving a higher and higher concentration of A in a step-by-step process known as *fractional distillation*. The process can be visualized by cobwebbing, in a way that is similar to the cobwebbing we did in solving the mean-field equation for magnets in Figure 5.12.

### Azeotropes

Up to this point, we've considered phase diagrams for binary liquid–vapor systems that have a "lens" shape, as in Figures 6.1–6.3. Other types of phase diagram are possible, however. The next step up in complexity is a phase

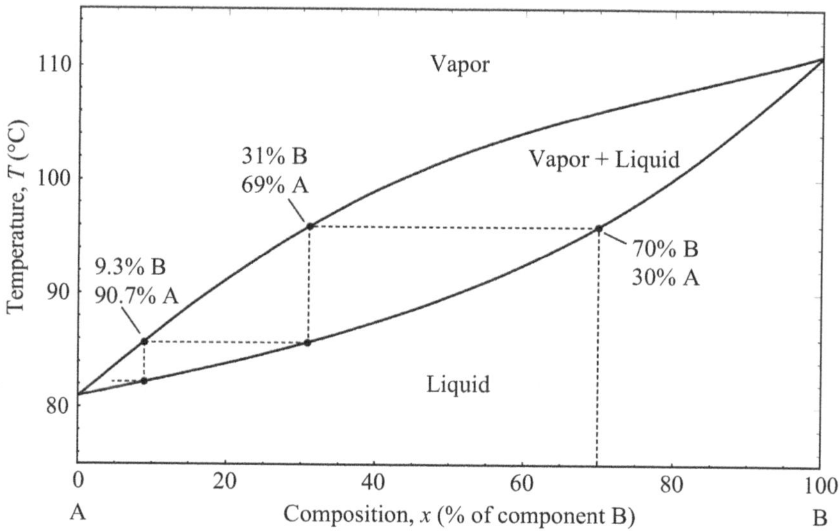


Figure 6.3 Successive distillations of an open binary mixture of A and B components. The mixture starts off with 30% A and ends with 90.7% A. The components are A = cyclohexane and B = toluene.

diagram where the lens pinches in to a single point at some intermediate composition. This pinch-off point, where the liquid and vapor have the same composition, is referred to as an *azeotrope*, which literally means "no (a-) change (trop-) on boiling (zee-)." We show examples in Figures 6.4 and 6.5.

First, in Figure 6.4, we see a situation in which the temperature of the azeotrope is greater than the boiling point of either pure substance. This is referred to as a *maximum azeotrope*. Cobwebbing in a case like this shows that fractional distillation results in the system moving away from the azeotrope in either direction. The result is a high concentration of A molecules if the starting composition is to the left of the azeotrope, and a high concentration of B molecules if the starting composition is to the right of the azeotrope.

In contrast, the azeotrope shown in Figure 6.5 has a lower temperature than the boiling point of either pure substance. This is referred to as a *minimum azeotrope*. Cobwebbing in this case shows that all starting values of the composition "flow" toward the azeotrope.

## 6.2 Solid–Liquid Phase Transitions

Continuing downward in temperature, we turn now to phase transitions from the liquid phase to the solid phase. There are a great many different types of

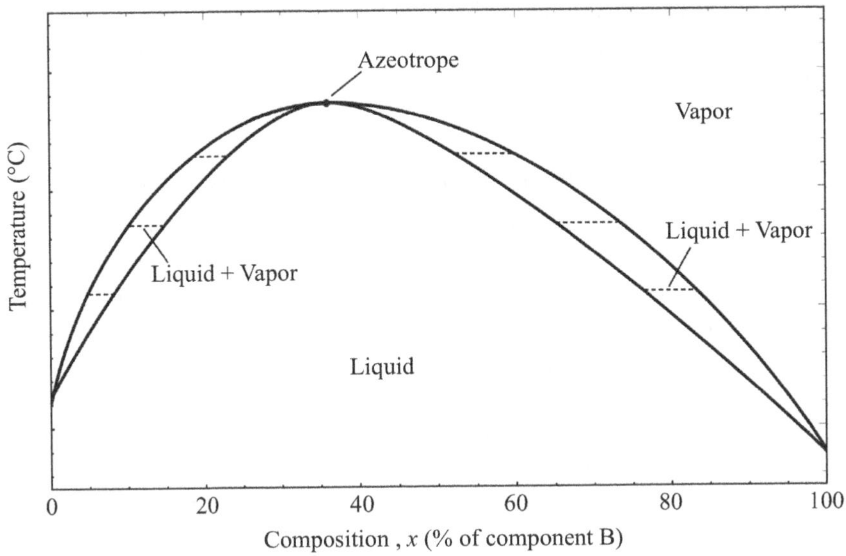


Figure 6.4 A temperature–composition phase diagram with a maximum azeotrope.

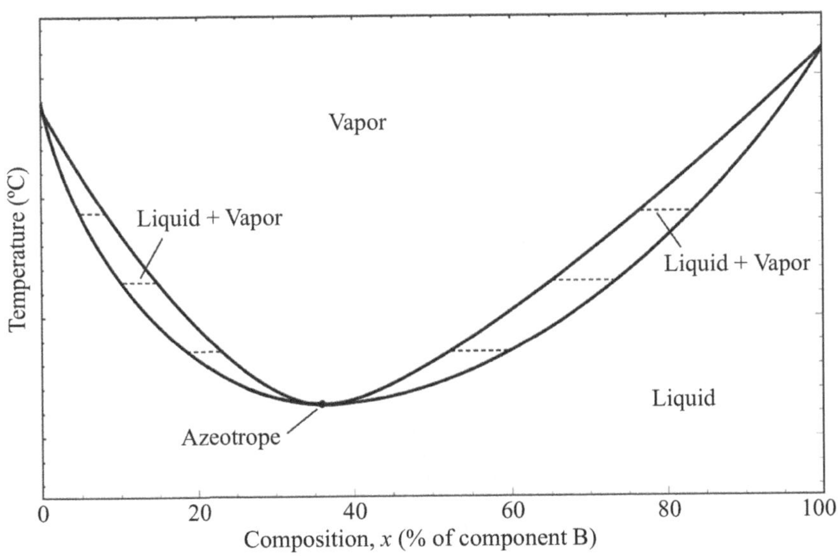


Figure 6.5 A temperature–composition phase diagram with a minimum azeotrope.

phase diagrams between these phases that occur in nature. In this section we discuss the most common varieties – other types are primarily variations on the themes presented here.

## Lens-Shaped, Single-Solid Phase Diagrams

The simplest liquid–solid phase diagram for a binary mixture is the lens-shaped temperature-versus-composition diagram shown in Figure 6.6. Notice that this is practically identical to the phase diagram for the liquid–vapor diagram shown in Figure 6.1. As with the liquid phase in Figure 6.1, we see that the solid phase is uniform throughout the composition range from pure A to pure B.

In Figure 6.6 we consider a mixture of two substances, labeled A and B. Substance A is an iron compound known as fayalite ($Fe_2SiO_4$), and substance B is a magnesium compound known as forsterlite ($Fe_2SiO_4$). The solid that is formed with any intermediate composition of A and B has an orthorhombic crystal structure, is greenish in color, and is referred to as olivine. The crystal structure works fine with either an iron or a magnesium atom at any given location on the lattice, thus allowing for the single, homogeneous solid phase.

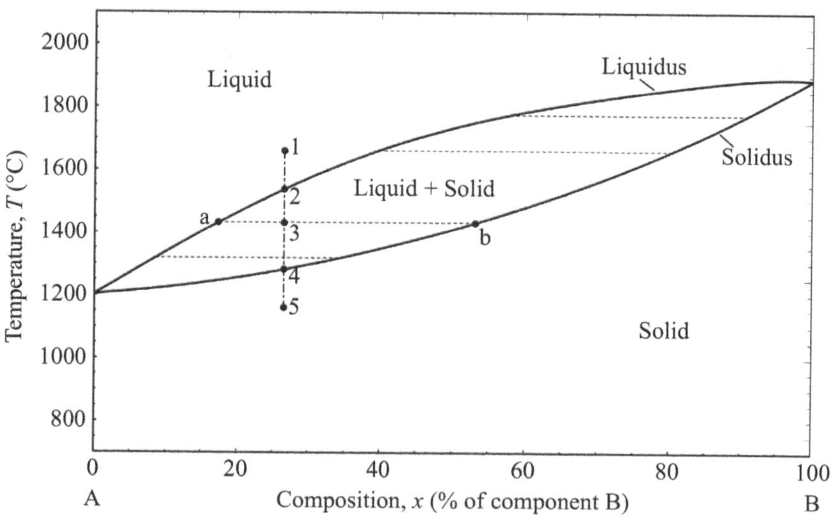


Figure 6.6 Lens-shaped, liquid–solid olivine phase diagram plotted with temperature $T$ on the vertical axis and composition $x$ (percentage of substance B) on the horizontal axis. The left side of the diagram is pure substance A, known as fayalite ($Fe_2SiO_4$), and the right side is pure substance B, known as forsterlite ($Mg_2SiO_4$).

To understand the phase diagram in Figure 6.6 more completely, consider the vertical dash-dotted line at a constant composition. This corresponds to a system that is prepared with approximately 72% A and 28% B. At point 1 the system is a homogeneous liquid. Lowering the temperature, the liquid begins to solidify at point 2. The upper curve of the "lens" is referred to as the *liquidus*. At point 3 the system is a coexistence of a liquid at the composition of point a, $x_a$, and a solid with the composition of point b, $x_b$. At point 4 the last drop of liquid has solidified, on the curve known as the *solidus*, and at point 5 the system is once again homogeneous, this time in the solid phase.

In the coexistence region the fractions of A, $f_A$, and B, $f_B$, are determined using the lever rule, as in Chapter 4. In the case shown in Figure 6.6 the distance from point 3 to point a (which is A-rich) is one-quarter the distance from a to b (which is B-rich). It follows that the system is three-quarters A and one-quarter B. In general, the lever rule can be expressed as follows:

$$f_{\mathrm{A}} = 1 - \left(\frac{x_3 - x_{\mathrm{a}}}{x_{\mathrm{b}} - x_{\mathrm{a}}}\right),$$

$$f_{\mathrm{B}} = 1 - \left(\frac{x_{\mathrm{b}} - x_3}{x_{\mathrm{b}} - x_{\mathrm{a}}}\right).$$

This result applies to any temperature in the range from point 2 to point 4.

A number of binary mixtures have lens-shaped, liquid–solid phase diagrams like this. Examples include copper and nickel, and gold and silver. In these cases, the solid can exist with either component in a given location. Not all solid structures can accommodate both substances, however, which leads to the coexistence of two distinct solid phases. We consider situations like this next.

## Two-Solid Phase Diagrams

It's quite common for binary mixtures to have different solid phases for the A and B components. This may be because the two components have distinctly different crystal structures in their solid phases, like simple cubic for A and face-centered cubic for B, for example. Alternatively, it may be because the size of the unit cell is different for the two substances, even though they have the same crystal structure. The latter is the situation for a mixture of NaI (sodium iodide) and RbI (rubidium iodide), as shown in Figure 6.7, both of which are cubic in their solid phases.

The left side of Figure 6.7 is the pure NaI system, and the right side is the pure RbI system. At high temperatures, plotted on the vertical axis, the system is in a homogeneous liquid phase. As the temperature is lowered, we encounter the liquidus curve on both the left and right sides of the diagram. On the left we

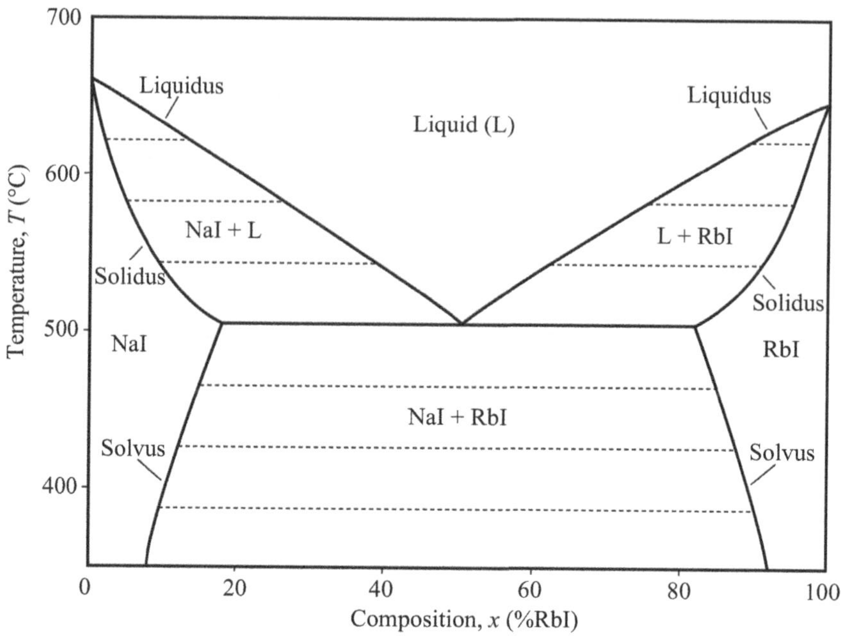


Figure 6.7 Liquid–solid phase diagram for a mixture of NaI (sodium iodide) and RbI (rubidium iodide).

have the liquid coexisting with the NaI solid, and on the right the liquid coexists with the RbI solid. Where the two liquidus curves meet we have a triple line, where three phases coexist. From left to right, these phases are NaI solid, liquid, and RbI solid.

At lower temperatures the system is completely solid. On the left the solid is NaI, on the right it is RbI, and in between we have a coexistence of these two solid phases. The curve separating the pure solid phase from the coexisting solid phases on either side of the phase diagram is referred to as the *solvus.*

In Figure 6.8 we show three different dash-dotted lines of constant composition. Each of these has distinctly different behavior as a function of temperature. We consider each in turn.

**Line 1–2–3.** On this line, the system starts out at point 1 as a high-temperature, homogeneous liquid. As the temperature is lowered, we cross the liquidus curve, and at point 2 the system consists of a coexistence of the liquid and the NaI solid phases. Lowering the temperature further to point 3, we cross the solidus curve and enter the pure NaI solid phase.

**Line 4–5–6.** At point 4, the system is a homogeneous high-temperature liquid. Crossing the liquidus curve we encounter a coexistence of the liquid

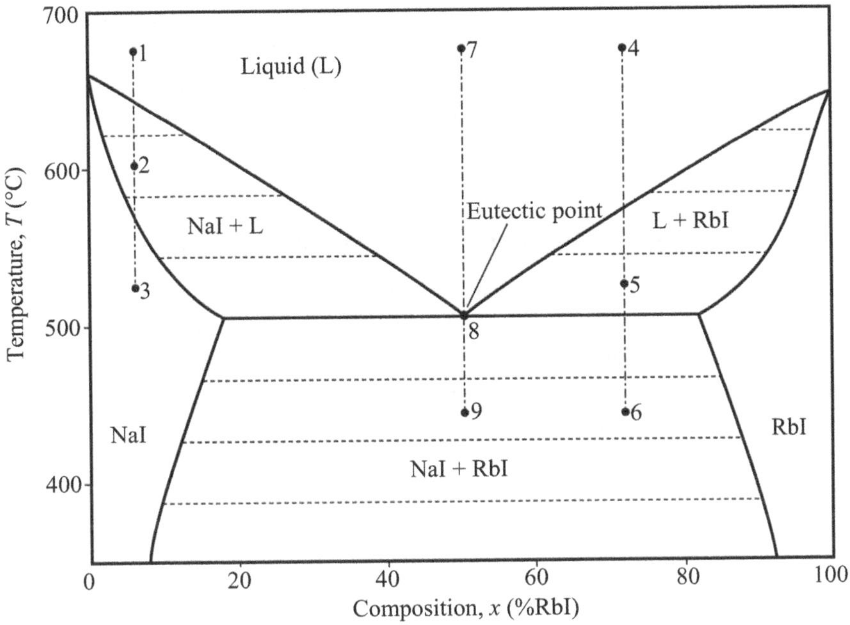


Figure 6.8 The liquid–solid phase diagram for NaI (sodium iodide) and RbI (rubidium iodide) with three different lines of constant composition illustrating distinct types of behavior.

and the RbI solid phases, as at point 5. At a lower temperature we cross the triple line and enter the region of coexistence between the NaI and RbI solid phases, as at point 6. We never encounter the RbI solid by itself.

**Line 7–8–9.** We save for last this special composition, where the liquid phase, as at point 7, extends to its lowest possible temperature, at point 8. This point is referred to as the *eutectic point*, or literally, the "good-solid point." When the liquid is cooled at this composition, as at point 9, it solidifies into two coexisting solid phases, with *no liquid left over*. It gives a good uniform solid phase. This composition is sometimes used in dentistry, where the binary compound used in fillings is mixed at this composition, so that when it solidifies the result is a uniform solid with no excess liquid.

## 6.3 Phase Transitions in Binary Liquid Mixtures

In this section we return to the liquid phase, which occurs at intermediate temperatures between the vapor and solid phases. As we shall see, additional types of phase transitions are possible within binary liquid mixtures. First, we consider the phase diagrams seen in many binary liquids, where the liquids transition between phases where the two components do or do not mix. In the next section, we extend our considerations to the rather surprising and counter-intuitive "upside-down" phase diagrams that are exhibited by some exceptional liquid mixtures.

## Miscibility/Immiscibility Phase Transitions

If we pour white cream into a cup of black coffee the two liquids mix completely, forming a homogeneous brownish liquid. In such a case, we say that the liquids are *miscible*, that is, the liquids are "mix-able." On the other hand, if we pour a quantity of water and oil into a container the liquids separate, with the water on the bottom and the less dense oil on top. These liquids are said to be *immiscible*, that is, they are "not mix-able."

This isn't the end of the story, however. The same two liquids can be miscible or immiscible depending on the temperature and composition of the mixture. This is illustrated in Figure 6.9.

For example, consider a test tube containing an equal amount of A and B liquids. At high temperature the system is a single, homogeneous fluid, as shown by the test tube at the top of the figure. As the temperature is lowered below the critical point, the liquids separate into an A-rich liquid and a B-rich liquid, as shown by the middle test tube. As the temperature is lowered further, the A-rich liquid becomes increasing rich in A molecules, and the B-rich liquid becomes increasing rich in B molecules. This is illustrated by the lowest test tube in the figure.

The physics behind Figure 6.9 can be understood as follows. First, at high temperature the system is random; that is, it is characterized by high entropy. Thus, the A and B molecules mix thoroughly throughout the system. In many systems, however, the energy of the system is lowered when like molecules are next to one another as compared to when opposite types of molecules are next to one another. As a result, at lower temperatures – where the system seeks to lower its energy – the A and B molecules segregate into different parts of the system, so that like molecules can gather together. As the temperature is lowered further, the A-rich and B-rich liquids become increasingly pure, as indicated near the bottom of Figure 6.9. This results in an additional decrease of the energy.

The net result is a dome-shaped coexistence curve separating the miscible liquid from the coexisting immiscible liquids. The region under the dome is sometimes referred to as a "miscibility gap." This means that it corresponds to a portion of the phase diagram where miscibility is no longer observed, that is,

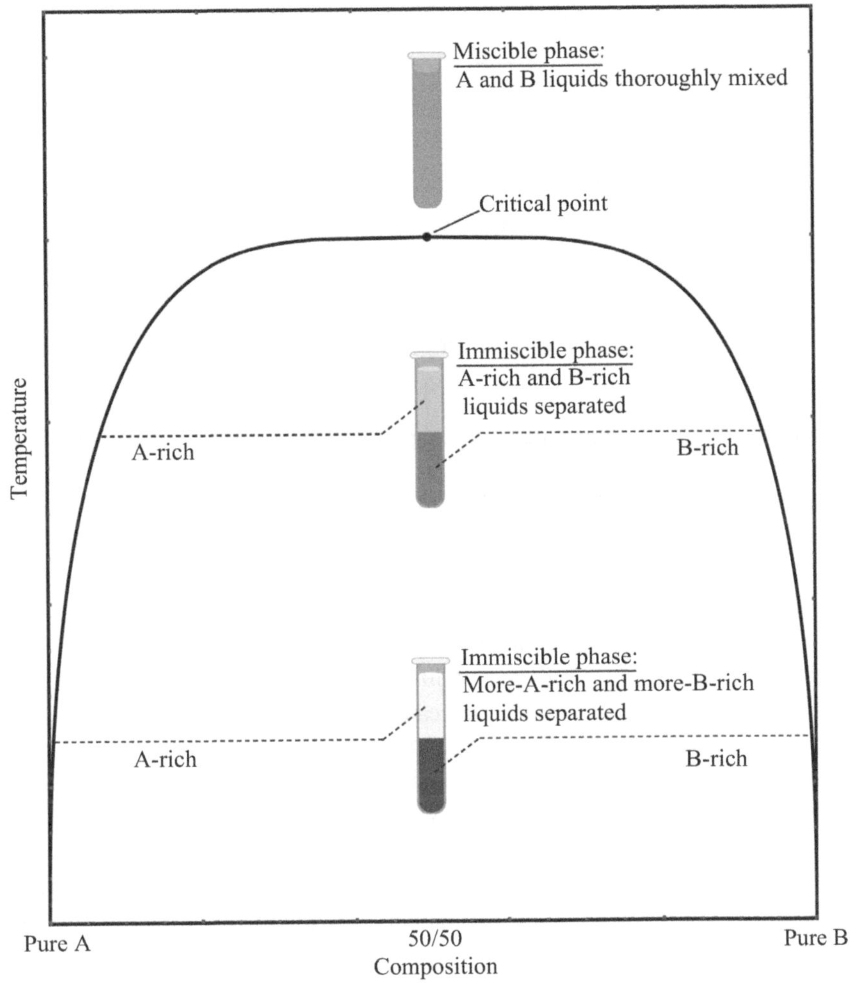


Figure 6.9 Miscible/immiscible phase diagram for a typical A–B liquid mixture. The system starts out as a single miscible liquid at high temperature, and becomes immiscible, with A-rich and B-rich liquids coexisting, as the temperature is lowered.

where one might say there's a "gap," or missing area, of miscibility. At lower temperatures, the liquids solidify, as discussed in the previous section.

## Ising Model for Binary Liquids

To undertake a theoretical study of phase diagrams like the one shown in Figure 6.9, we can use the Ising model. That may seem surprising at first, given

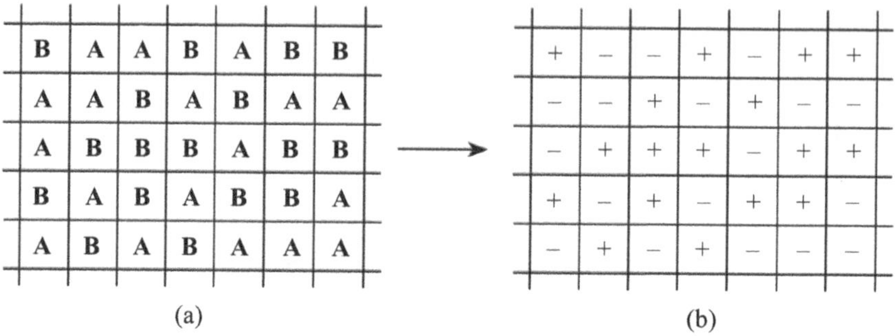


Figure 6.10 Making a connection between (a) a binary liquid mixture and (b) the corresponding Ising magnet.

that this model was introduced in Chapter 5 as a model for magnetism. How can a model for magnetic metals apply to a binary mixture of liquids? It may seem odd, but the wide applicability of the Ising model is one of the reasons it has been such an important part of theoretical physics for many decades.

To draw a connection between binary liquid mixtures and metallic magnets, consider Figure 6.10. In panel (a) we see a binary liquid mixture of A and B molecules arranged on an atomic-scale lattice. We show a two-dimensional lattice, but that is just for ease of illustration – the actual system is three-dimensional. Each site of the lattice can hold a single molecule.

In Figure 6.10(b) we show a corresponding lattice of magnetic dipoles, as in the Ising model. These dipoles can point up, as indicated with the "+" sign, or they can point down, as indicated with the "–" sign. In this example, we associate a "–" sign with A molecules and a "+" sign with B molecules. It follows that the interactions and phase transitions in an Ising model can be identified with corresponding interactions and phase transitions in a binary liquid mixture.

To be more specific about this connection, consider an Ising model with a nearest-neighbor interaction given by $K_{\text{Ising}}$. The reduced Hamiltonian in this case is

$$-\beta H_{\text{Ising}} = \sum_{\langle i,j \rangle} K_{\text{Ising}} s_i s_j.$$

In this expression, the sum is over nearest-neighbor pairs on the lattice, and the variables on each site are $s_i = \pm 1$.

In Figure 6.11(a) we show the energy levels for the Ising model. Notice that the reduced interaction energy is $K_{\text{Ising}}$ if nearest-neighbor spins point in the same direction, $s_i = s_j$, and is $-K_{\text{Ising}}$ if nearest-neighbor spins point in opposite directions, $s_i \neq s_j$.

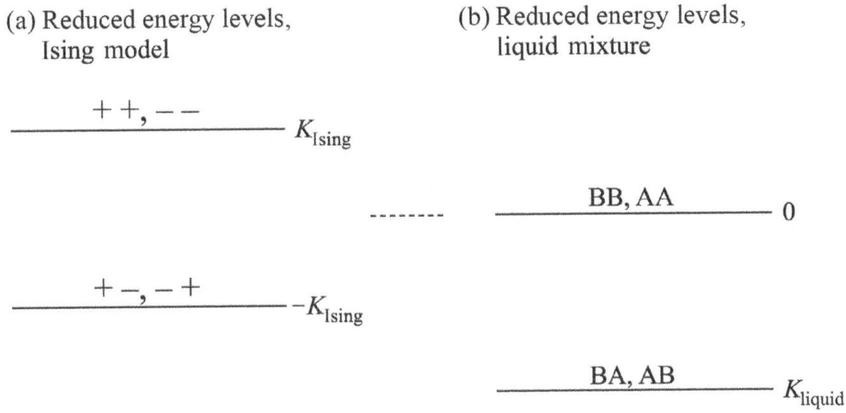


Figure 6.11 Energy levels for the (a) Ising model and (b) binary liquid mixtures.

To model the binary liquid mixture, we consider nearest-neighbor interactions between the A and B molecules. Since A and B don't have numerical values, for example, +1 and –1, we can't just multiply As and Bs together, as we do with the $s_i$ and $s_j$ variables in the Ising model. Instead, we can get numerical values in our reduced Hamiltonian by using the Kronecker delta function, $\delta$. Recall that

$$\delta_{s_i s_j} = \begin{cases} 1 & \text{if } s_i = s_j, \\ 0 & \text{if } s_i \neq s_j. \end{cases}$$

With this in mind, we can write a reduced Hamiltonian for the binary liquid system as

$$-\beta H_{\text{Liquid}} = \sum_{\langle i,j \rangle} K_{\text{Liquid}} \left(1 - \delta_{s_i s_j}\right). \tag{6.1}$$

In this Hamiltonian, the sum is again over nearest-neighbor pairs, and the variables at each site of the lattice take on two values: $s_i = \text{A, B}$.

The energy levels produced by this reduced Hamiltonian are shown in Figure 6.11(b). When $s_i = s_j$ (BB and AA), the reduced energy is zero – this defines the zero of energy for the system. On the other hand, when $s_i \neq s_j$ (BA and AB), the reduced energy is $K_{\text{Liquid}}$. If we want the reduced energy to be favorable when like molecules are next to one another, as opposed to when unlike molecules are neighbors, it follows that unlike-molecule neighbors must have a negative reduced energy, $K_{\text{Liquid}} < 0$, as indicated in Figure 6.11.

To connect the Ising model energy levels to the liquid mixture energy levels, note that the difference in reduced energy between the two levels for the Ising model is $K_{\text{Ising}} - (-K_{\text{Ising}}) = 2K_{\text{Ising}}$. The corresponding difference in reduced energy for the liquid mixture is $0 - K_{\text{Liquid}} = -K_{\text{Liquid}}$. Setting these quantities equal to one another, we see that the connection between the two models is

$$K_{\text{Ising}} = -\frac{1}{2}K_{\text{Liquid}}.$$

This allows us to make a specific, numerical connection between the two systems.

To explore this connection in more detail, we introduce a new quantity to play the role of temperature. Recall that reduced interactions, such as $K_{\text{Liquid}}$, have the following form:

$$|K_{\text{Liquid}}| = \frac{|\text{interaction energy}|}{k_B T}.$$

The absolute value signs are introduced here to make it clear that $K_{\text{Liquid}}$ and the interaction energy are negative. Absorbing the interaction energy and Boltzmann's constant, $k_B$, into the temperature, we can simplify this expression with a dimensionless *reduced temperature*, $T_r$, defined as

$$T_r = 1/|K_{\text{liquid}}|.$$

With this definition, we have the following connection between $K_{\text{Ising}}$ and the reduced temperature:

$$K_{\text{Ising}} = \frac{1}{2T_r}.$$

We plot $K_{\text{Ising}}$ as a function of $T_r$ in Figure 6.12.

Note that as the reduced temperature is lowered, the value of $K_{\text{Ising}}$ increases. At a certain point it crosses the critical value for the Ising model, which for the three-dimensional simple cubic model is given by

$$K_c = 0.22165\ldots$$

This value is indicated by the horizontal dashed line in Figure 6.12. Above this dashed line the system is in the immiscible, two-liquid phase, with A molecules on one side of the container and B molecules on the other side. The critical point in Figure 6.12 corresponds to the critical point at the highest point of the coexistence dome in Figure 6.9, and corresponds to the following value of the reduced temperature:

$$T_r = \frac{1}{2K_c} = 2.25581\ldots$$

This is the value shown in Figure 6.12 along the horizontal axis.

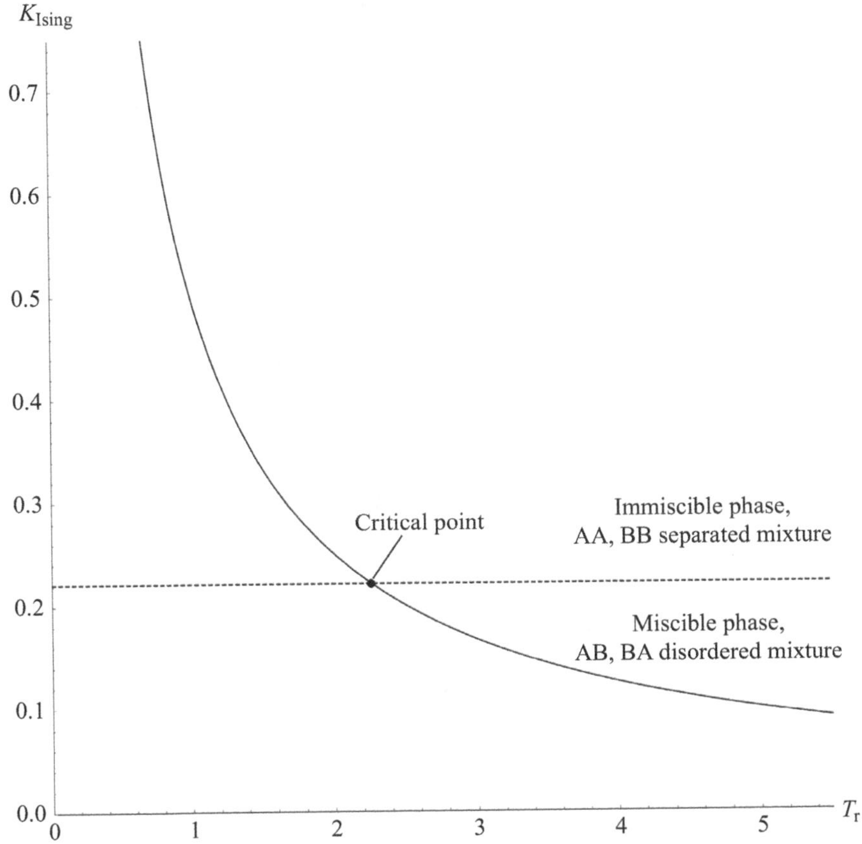


Figure 6.12 The Ising coupling $K_{\text{Ising}}$ as a function of the reduced temperature $T_{\text{r}}$. The critical point occurs when $K_{\text{Ising}} = K_{\text{c}} = 0.22165\ldots$, and corresponds to the critical point at the top of the coexistence dome in Figure 6.9. The value of the reduced temperature at the critical point is $T_{\text{r}} = 2.25581\ldots$

One last connection to make between these two models is to relate the composition of the liquid mixture, $x$, to the magnetization of the Ising model, $m$. First, we let $s_i = -1$ correspond to $s_i = A$, and $s_i = +1$ to correspond to $s_i = B$. Now, if $x$ measures the fraction of B in a mixture, then $x = 0$ corresponds to all A (and $m = -1$), and $x = 1$ corresponds to all B (and $m = +1$). We can quantify this with a linear relationship between $x$ and $m$ as follows:

$$x = a + b\,m.$$

In this expression, $a$ and $b$ are constants to be determined. We can find these constants using the following two conditions:

$$0 = a + b(-1),$$
$$1 = a + b(+1).$$

Solving for $a$ and $b$ yields

$$a = b = \frac{1}{2}.$$

Thus, the relation between $x$ and $m$ can be written as

$$x = \frac{1}{2}(1 + m). \tag{6.2}$$

In what follows we will speak in terms of $x$ or $m$, depending on the situation. We now focus on determining the Ising magnetization.

## Ising Magnetization

To determine the Ising magnetization in three dimensions, we start with the ground state at zero temperature – all spins up or all spins down – and consider excitations as the temperature is raised. The all-spins-up ground state is shown in Figure 6.13(a) for the case of two dimensions, though we are actually interested in three dimensions. The lowest-energy excitation from the ground state is shown in Figure 6.12(b), where a single spin is flipped to the down direction.

Notice that the lowest-energy excitation "breaks" bonds with the nearest neighbors of the flipped spin. The broken bonds come in pairs – one pair to the left and right, one pair up and down, and, for the three-dimensional lattice, one pair into and out of the page. A pair of broken bonds changes the reduced

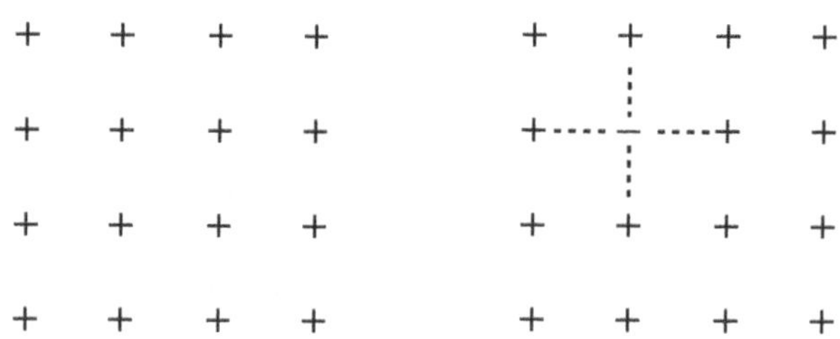

(a) Ground state

(b) First excited state

Figure 6.13 (a) The ground state, with all spins in the same direction, and (b) the first excited state, with one spin flipped, for an Ising model on a square lattice.

energy from $2K$ to $-2K$, for a change of $-4K$, and thus the Boltzmann factor associated with this change is

$$y = e^{-4K}.$$

For this first excitation in three dimensions, the probability that it occurs is proportional to the Boltzmann factor coming from three different pairs of bonds, one for each coordinate direction. Thus, the probability for the excitation is proportional to

$$y^3 = e^{-12K}.$$

Note that this excitation can occur anywhere on the lattice, which we assume has $N$ sites, where ultimately $N$ goes to infinity.

The magnetization, which is the average value of a spin, can be written as

$$m = \langle s_i \rangle = \frac{1}{N} \frac{\sum\limits_{\text{all states}} \left( \sum\limits_i s_i \right) e^{-\beta H}}{\sum\limits_{\text{all states}} e^{-\beta H}}. \tag{6.3}$$

In this expression, the denominator is the sum of Boltzmann factors for all states. This is the partition function, as defined in Eq. (5.2). As we saw there, dividing by the partition function converts the Boltzmann factors in the numerator into probabilities. We then multiply the probabilities by the sum over all spins on the lattice. Since there are $N$ sites, and all the spins are identical, we divide the result by $N$ to obtain the average value for a single spin.

To obtain a series expansion for the magnetization in powers of $y$, we start by writing Eq. (6.3) for the all-spins-up ground state and the $N$ first excited states. The result is

$$m = \frac{1}{N} \left( \frac{Ne^{3NK} + N[(N-1)(+1) + (1)(-1)]e^{3NK}e^{-12K}}{e^{3NK} + Ne^{3NK}e^{-12K}} \right). \tag{6.4}$$

We now explore this expression in detail.

The first term in the denominator, $e^{3NK}$, represents the fact that when all spins are pointing up, the reduced energy of the lattice is $3NK$; that is, three nearest neighbors per site (to avoid overcounting) with reduced energy $K$. The second term in the denominator represents the $N$ first excited states, with one flipped spin and a reduced energy of $3NK - 12K$.

The first term in the numerator, $Ne^{3NK}$, represents the ground state Boltzmann factor multiplied by the sum of all $N$ spins on the lattice – all of which are up. The second term in the numerator represents the excitation with one down spin. This excitation can occur at any of the $N$ sites on the lattice, and has a Boltzmann factor of $e^{3NK}e^{-12K}$. The sum over all spins for this state is $(N-1)(+1)$ for the up spins, plus $(1)(-1)$ for the one down spin, for a total of $N-2$.

Simplifying Eq. (6.4) by dividing through by $N$ and canceling factors of $e^{3NK}$, we find

$$m = \left(\frac{1+(N-2)e^{-12K}}{1+Ne^{-12K}}\right).$$

Expanding the denominator, and replacing the exponentials with powers of $y$, we find

$$m = \left(1+(N-2)y^3\right)\left(1-Ny^3+\ldots\right) = 1-2y^3-\ldots.$$

This result, which is the lowest-order change to the ground state, $m = 1$, is shown by the top dashed curve in Figure 6.14.

We can understand this result as follows. First, the ground state corresponds to a magnetization of 1, as indicated by the first term in the expression. Next, the first excited state occurs with a probability proportional to $y^3$, and

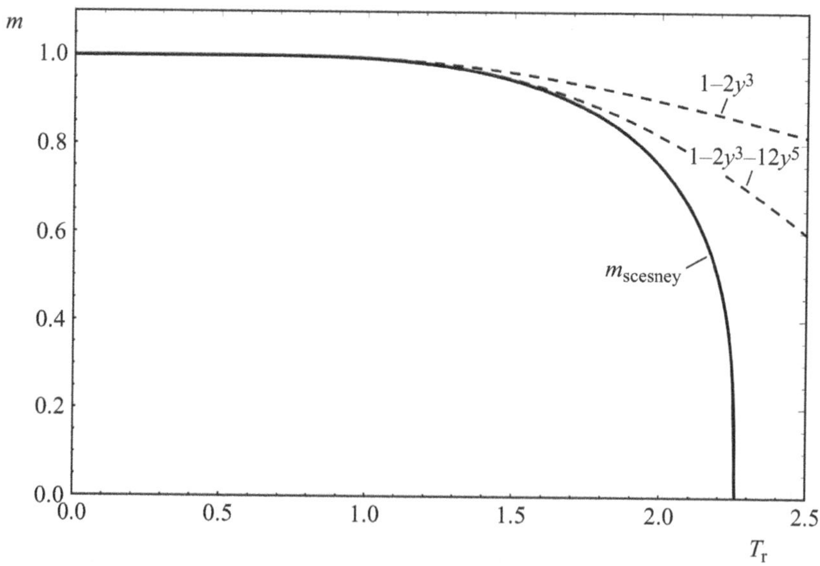


Figure 6.14 Ising magnetization for the simple cubic lattice. The dashed curves give the first two corrections to the ground state magnetization, and the solid curve is the best available result for the magnetization, $m_{scesney}$, as given in Eq. (6.5).

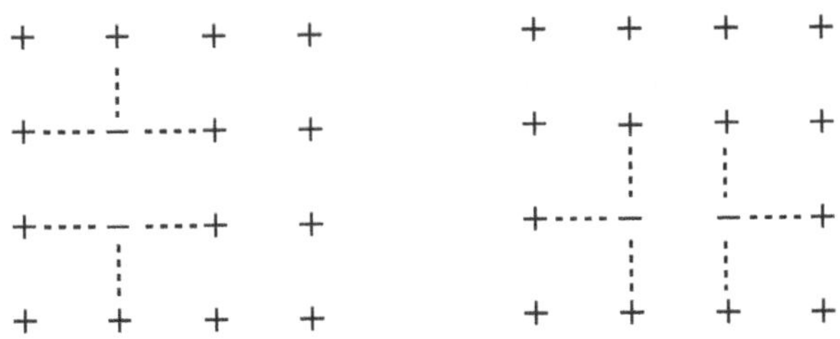

(a) Second excited state, vertical orientation

(b) Second excited state, horizontal orientation

Figure 6.15 The second excited state for the Ising model on a square lattice. This excitation has two neighboring flipped spins, oriented either (a) vertically or (b) horizontally.

corresponds to a change in spin of –2, from +1 to –1. Thus, we have $m = 1 - 2y^3 - \cdots$.

We show the second excited state in Figure 6.15. Here, two spins that are nearest neighbors are flipped. This excitation can occur vertically, as in Figure 6.15(a), horizontally, as in Figure 6.15(b), or out of the page (not shown).

There are five pairs of broken bonds for each of these excitations. As an example, in Figure 6.15(a) we see two pairs that are to the left and right, one pair that is up and down, and two pairs that go into and out of the page. Thus, the probability of this excitation is proportional to $y^5$. The change from flipping the spins is –2 for each spin, that is, –4 for each nearest-neighbor pair. There are three nearest-neighbor pairs – one for each coordinate direction – for a change in spin of –12. Thus, the magnetization through second-order can be written as

$$m \approx 1 - 2y^3 - 12y^5 - \dots .$$

This expression is shown by the lower dashed curve in Figure 6.14.

Continuing with this process, it is possible to find a number of exact terms in the expansion of the magnetization – there is no closed-form solution known for three dimensions, so exact series expansions are the only option for exact results. Counting the contributions to higher-order terms quickly becomes quite difficult. In fact, there is a rule of thumb that states that "the amount of work required to obtain the next term in the series expansion is equal to the amount of work required to obtain all of the previous terms combined." Stated another way, the amount of work required to calculate terms in a series expansion grows exponentially.

The largest number of terms in the series expansion for the Ising model on the simple cubic lattice was obtained by Scesney (1970). He produced the following expression for the magnetization:

$$m_{\text{scesney}} = \left(1 - \frac{y}{y_c}\right)^{5/16} \left(\frac{Num}{Den}\right). \tag{6.5}$$

In this expression,

$$y_c = e^{-4K_c} = 0.412047\ldots$$

In addition, the numerator (*Num*) and denominator (*Den*) in the expression are as follows:

$$\begin{aligned} Num &= 1 + 3.20844y - 6.85896y^2 - 19.4368y^3 + 8.91501y^4 + 23.8551y^5 \\ &\quad + 52.0062y^6 - 9.11569y^7, \\ Den &= 1 + 2.44985y - 9.92584y^2 - 15.1289y^3 + 28.7181y^4 + 20.4259y^5 \\ &\quad + 26.9927y^6 - 57.4305y^7. \end{aligned}$$

We show $m_{\text{scesney}}$ in Figure 6.14 with the solid curve.

If we expand $m_{\text{scesney}}$ about zero temperature, that is about $y = 0$, we find

$$m_{\text{scesney}} \approx 1 - 2y^3 - 12y^5 - \ldots.$$

Thus, we see that the terms we've calculated by hand are included in $m_{\text{scesny}}$. In addition, all of the additional terms obtained by Scesney are included as well. In fact, the Scesney magnetization can be considered as an exact result for all practical purposes, and gives results that are correct within the thickness of the curve in a plot.

## Comparing with Experiments

In Figure 6.14 we plot the magnetization on the vertical axis and temperature on the horizontal axis. We plot only positive values of the magnetization in this figure, but it should be noted that we could just as well have started with the all-spins-down ground state, and obtained a reflected version of the plot. In what follows, we will include both the positive and negative branches of the magnetization.

In addition, to compare with experiments we rotate the plot in Figure 6.14 so that temperature is on the vertical axis, and magnetization is on the horizontal

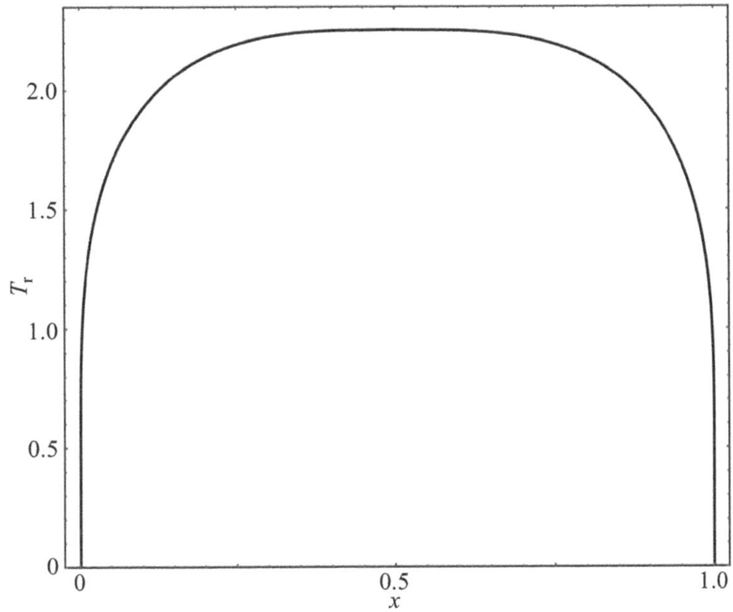


Figure 6.16 Reduced temperature $T_r$ versus composition $x$, using the Scesney result for the three-dimensional Ising magnetization, $m_{scesney}$.

axis. We then convert the magnetization $m$ to the composition $x$ using Eq. (6.2). The result is shown in Figure 6.16. This shows a typical dome-shaped coexistence curve, as discussed in Figure 6.9, where the region within the dome corresponds to two liquids in coexistence.

To compare with a specific experiment, we consider glycerol + methyl ethyl ketone. Reducing the temperature by dividing by the critical temperature of the mixture, and plotting versus composition, we obtain the plot shown in Figure 6.17.

The dots in Figure 6.17 represent experimental results for the binary liquid mixture. The solid curve is the Scesney magnetization. Clearly, the agreement between a liquid mixture and the Ising model is quite satisfactory.

## 6.4 Closed-Loop Phase Diagrams

Most binary liquid mixtures have dome-shaped phase diagrams, like the one shown in Figure 6.17. There are exceptional binary mixtures, however, that have what are known as *closed-loop phase diagrams*. An example, based on the mixture glycerol + m-toluidine, is shown in Figure 6.18.

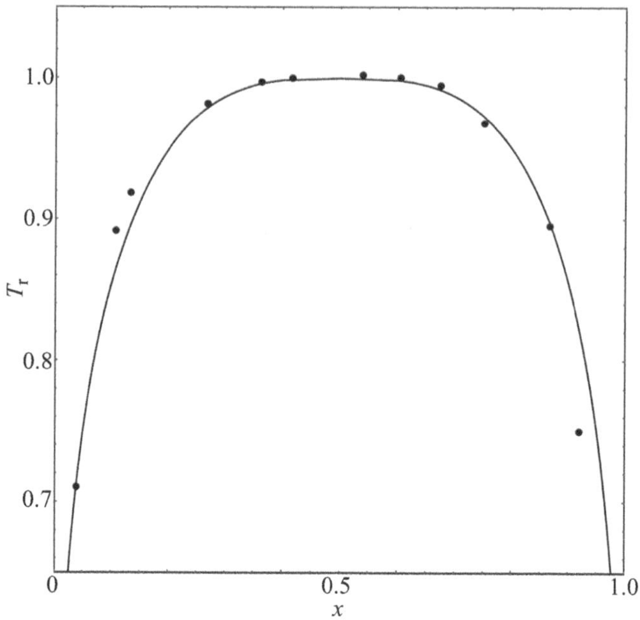


Figure 6.17 Phase diagram for glycerol + methyl ethyl ketone. The dots give experimental results, and the solid curve is the portion of $m_{scesney}$ near the peak of Figure 6.16.

The top of the closed loop is similar to the top of a normal dome-shaped coexistence curve for a mixture of A and B molecules. At temperatures above the loop the system is in the miscible phase, with the A and B liquids thoroughly mixed together. This is shown in the top test tube in Figure 6.18, which, like the other test tubes in the figure, is for the case of equal amounts of A and B liquids. As the temperature is lowered, the system passes through an upper critical point at the temperature $T_U$ and separates into A-rich and B-rich liquids, as in the middle test tube in the figure. Finally, as the temperature is lowered further, the system passes through the lower critical point, at the temperature $T_L$, and the A and B liquids become miscible once again. This is shown by the lower test tube in Figure 6.18.

The phase transition at the lower critical point is from an immiscible phase at high temperature to a miscible phase at low temperature. This is an "upside-down" phase transition compared to what is seen in normal binary mixtures. In fact, a phase transition like this is puzzling, because its high-temperature phase *appears* to be more ordered than its low-temperature phase – just the opposite of normal phase transitions.

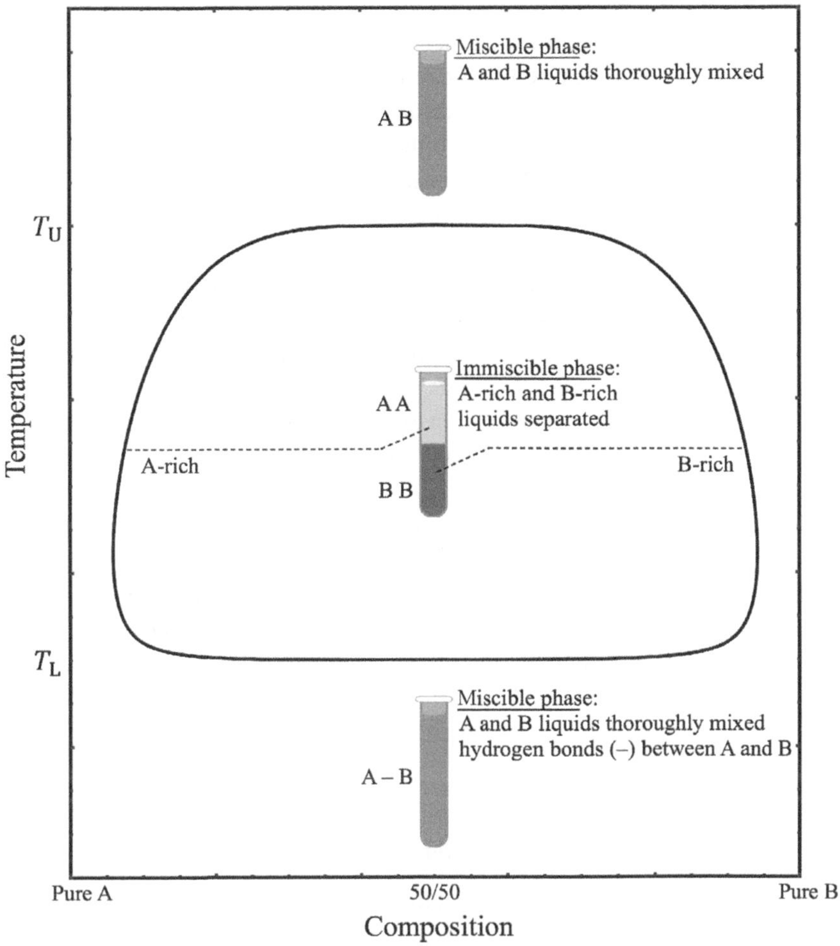


Figure 6.18 The features of a closed-loop phase diagram. This phase diagram is based on glycerol + m-toluidine, and corresponds to $R = 10.5$ and $q = 8000$ in the WV model, introduced later in this section.

## The Importance of Hydrogen Bonding

To understand these puzzling phase diagrams, we need to take account of the fact that mixtures that show closed-loop phase diagrams have A and B molecules that are capable of forming hydrogen bonds with one another. This is important because hydrogen bonds are energetically favorable, but entropically unfavorable. What we mean by "entropically unfavorable" is that when a hydrogen bond forms between A and B molecules the relative orientation of the molecules is constrained to within a very small range. As we shall see, this has the effect of reducing the entropy of the system.

To see how hydrogen bonding can produce an upside-down phase transition, we need to consider two very different types of entropy that are at play in these systems. One type of entropy is the "mixing entropy" associated with the combining of the A and B molecules. Above the lower critical point, the immiscible liquids have low mixing entropy because the A and B molecules are separated into different parts of the system. Below the lower critical point, however, the mixing entropy increases, because the A and B molecules are now mixed together – so it seems entropy has *increased* with *decreasing* temperature, which is just the opposite of what happens in a system that is thermodynamically stable.

This isn't the end of the story, however. There's another contribution to the entropy, referred to as "orientational entropy," which is associated with the orientation of the molecules. When A and B molecules are unbonded, they can orient freely in practically any direction, and thus they have high orientational entropy. When they are hydrogen bonded, however, their orientational entropy drops significantly because of the tight restriction on their relative orientation. The *reduction in orientational* entropy *more than compensates for the increase in mixing* entropy, and hence the *total entropy* of the system *decreases* with decreasing temperature, as it must. This is the basic physical mechanism behind the closed-loop phase diagrams.

## The WV Model for Closed Loops

We now develop a theoretical model that incorporates the physical mechanism of hydrogen bonding in a binary liquid mixture. First, to represent the two components of the system, we again use an Ising variable, $s_i = \mathrm{A}, \mathrm{B}$, for each site of the lattice. Next, we want a variable to represent the orientation of a molecule. We assign the variable $\sigma_i = 1, 2, 3, \ldots, q$, known as a *Potts variable*, to each site of the lattice. The $q$ different values of $\sigma_i$ correspond to $q$ different orientations of the molecule, in contrast to the mean-field theory of the previous chapter where $q$ represented the number of nearest neighbors.

The full model will combine effects from both of these variables. Specifically, we shall say that a hydrogen bond occurs between nearest-neighbor molecules on the lattice when the following two conditions are satisfied: (1) The neighboring molecules are A and B molecules, that is, $s_i \neq s_j$; (2) The neighboring molecules have the same value of their relative orientation, that is, $\sigma_i = \sigma_j$.

A theoretical model with these characteristics was introduced by Walker and Vause in 1980. It is referred to as the *WV model*, and its reduced Hamiltonian has the form

$$-\beta H_{\mathrm{WV}} = \sum_{\langle i,j\rangle} \left[K_1\left(1-\delta_{s_is_j}\right)\delta_{\sigma_i\sigma_j} + K_2\left(1-\delta_{s_is_j}\right)\left(1-\delta_{\sigma_i\sigma_j}\right)\right]. \tag{6.6}$$

Notice that there are two reduced interaction energies in the WV model, $K_1$ and $K_2$. Both of these reduced energies occur when $s_i \neq s_j$ (so that $1-\delta_{s_is_j}=1$), that is, when nearest neighbors are A B or B A. The favorable reduced energy $K_1 > 0$ occurs when $\sigma_i = \sigma_j$, and hence it represents the bonding energy between A and B molecules. The unfavorable reduced energy $K_2 < 0$ occurs when $\sigma_i \neq \sigma_j$, that is, it represents the nonbonding interaction between A and B molecules. Zero energy corresponds to like molecules, A A or B B, as nearest neighbors, regardless of their orientations. All of these energies are summarized in the reduced energy-level diagram in Figure 6.19.

Also shown in Figure 6.19 are the degeneracies – or number of ways of occurring – for each reduced energy level. Notice that the degeneracy for the bonding level, $K_1$, is of order $q$, whereas the degeneracy for the nonbonding levels, 0 and $K_2$, are of the order $q^2$. Because hydrogen bonds are highly

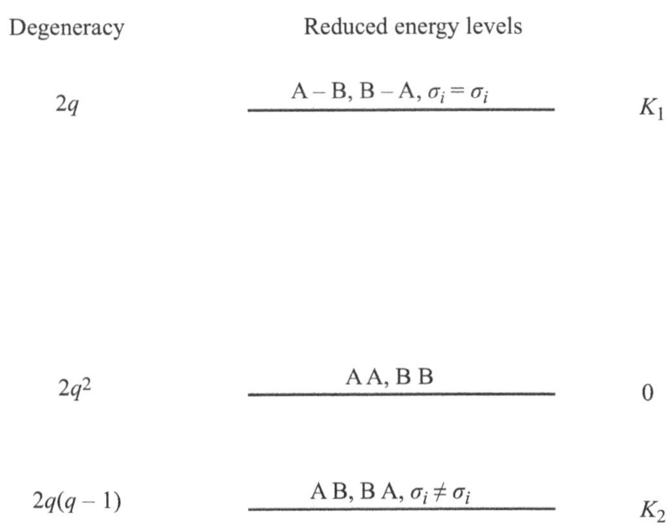


Figure 6.19 Reduced energy levels for the WV model with $-K_1/K_2 = 2.5$. Bonding is indicated by A – B and B – A. Note that the bonding level has a favorable energy and has a degeneracy smaller than that of the nonbonding levels by a factor of $q$.

directional, occurring over only a small range of angles, it follows that $q$ will have a rather large value; that is, a hydrogen bond can point in a very large number, $q$, of different directions. In fact, as we shall see shortly, the value of $q$ that corresponds well with experiment is in the ballpark of $q = 8000$. As a result, the degeneracies of the nonbonding levels are *much* greater than the degeneracy of the bonding level.

At high temperature, the system spends roughly equal time in all configurations, and hence only a small portion of time is spent in the bonding configurations. It follows that the levels 0 and $K_2$ dominate the system, and the bonding level has little impact. The nonbonding levels are basically the same as those for the Ising model in Figure 6.11, with $K_2$ playing the role of $K_{\text{Liquid}}$. As a result, the high-temperature behavior of the WV model is essentially the same as that of the Ising model, resulting in the upper critical point in Figure 6.18.

On the other hand, when the temperature is lowered, the system preferentially spends time in the most favorable energy configurations. In the WV model, this means that the $K_1$ level now begins to dominate the system, resulting in A and B molecules coming together and forming hydrogen bonds. This results in the lower critical point in Figure 6.11, and the completion of the closed loop.

We now turn to a method of analysis for the WV model that relates it to the Ising model and allows us to tie into the great body of knowledge associated with that model.

## The Partial-Summation Transformation

In Figure 6.20, we show the partial-summation transformation for the WV model, which was introduced by Walker and Goldstein in 1982. The basic idea behind this transformation is that by taking account of the orientational degrees of freedom in the WV model, we can derive an effective coupling between Ising variables that includes the effects of hydrogen bonding.

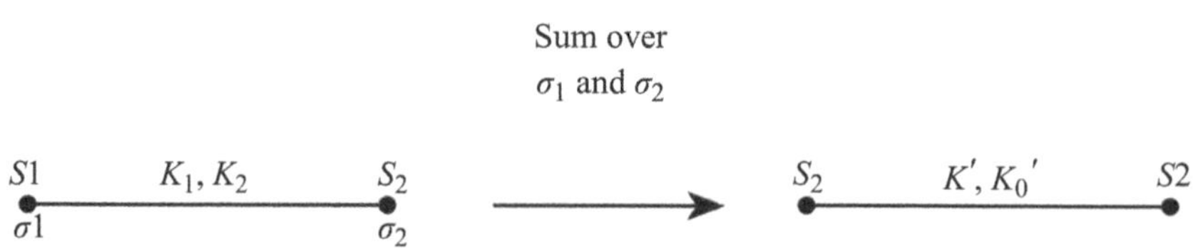


Figure 6.20 Partial summation from the WV model to the Ising model.

To implement the partial-summation transformation, consider a single bond of the WV model connecting sites 1 and 2. This bond is representative of all the bonds in the system. Next, sum over the $\sigma$ variables on the bond, and generate an effective coupling between the remaining $s$ variables. This transformation is exact – the only approximation is that we consider a single bond rather than the entire lattice.

To begin, write the reduced Hamiltonian for the WV model for the two sites of the single bond. The result is

$$-\beta H_{\mathrm{WV},12} = K_1(1-\delta_{s_1 s_2})\delta_{\sigma_1\sigma_2} + K_2(1-\delta_{s_1 s_2})(1-\delta_{\sigma_1\sigma_2}).$$

The effective reduced Hamiltonian after summing over the $\sigma$ variables can be written as

$$-\beta H'_{12} = K'(1-\delta_{s_1 s_2}) + K'_0.$$

In this expression, $K'$ plays the role of $K_{\mathrm{Liquid}}$ in Figure 6.11, and $K'_0$ is a shift in the zero level produced by the transformation.

To implement the transformation, we require that the sum over all Boltzmann factors in the effective model be equal to the sum over all Boltzmann factors in the WV Ising model. Specifically,

$$\sum_{\{s_1,s_2\}} e^{-\beta H'_{12}} = \sum_{\{s_1,s_2,\sigma_1,\sigma_2\}} e^{-\beta H_{\mathrm{WV},12}}. \tag{6.7}$$

This gives an exact connection between $-\beta H'_{12}$ and $-\beta H_{\mathrm{WV},12}$.

The best way to proceed is to consider special cases for the values of $s_1$ and $s_2$. First, we assign $s_1$ = A and $s_2$ = B. The result for the left-hand side of Eq. (6.7) is

$$e^{-\beta H'_{12}(s_1=\mathrm{A},s_2=\mathrm{A})} = e^{K'_0}.$$

For the right-hand side of Eq. (6.7) we have

$$\sum_{\{\sigma_1,\sigma_2\}} e^{-\beta H_{\mathrm{WV},12}(s_1=\mathrm{A},s_2=\mathrm{A})} = \sum_{\{\sigma_1,\sigma_2\}} 1 = q^2.$$

Equating the two sides of the equation yields

$$e^{K'_0} = q^2.$$

This determines one of the variables in the effective reduced Hamiltonian.

To determine the other variable, $K'$, we consider the special case where $s_1$ = A and $s_2$ = B. The left-hand side of Eq. (6.7) is now

$$e^{-\beta H'_{12}(s_1=\mathrm{A},s_2=\mathrm{B})} = e^{K'+K'_0}.$$

For the right-hand side we have

$$\sum_{\{\sigma_1,\sigma_2\}} e^{-\beta H_{\mathrm{WV},12}(s_1=\mathrm{A},s_2=\mathrm{B})} = \sum_{\{\sigma_1,\sigma_2\}} e^{K_1\delta_{\sigma_1\sigma_2}+K_2\left(1-\delta_{\sigma_1\sigma_2}\right)}.$$

Carrying out the sum over the $\sigma$ variables yields

$$\sum_{\{\sigma_1,\sigma_2\}} e^{K_1\delta_{\sigma_1\sigma_2}+K_2\left(1-\delta_{\sigma_1\sigma_2}\right)} = qe^{K_1} + q(q-1)e^{K_2}.$$

In this sum, notice that there are $q$ ways for the $\sigma$ variables to be equal to one another, with the Boltzmann factor $e^{K_1}$, and $q(q-1)$ ways for the $\sigma$ variables to have different values, with the Boltzmann factor $e^{K_2}$. Equating the two sides of the equation yields

$$e^{K'+K'_0} = qe^{K_1} + q(q-1)e^{K_2}.$$

Recalling that $e^{K'_0} = q^2$, and taking the log of both sides of this equation, we find

$$K' = \ln\left[\frac{qe^{K_1} + q(q-1)e^{K_2}}{q^2}\right] = \ln\left[\frac{e^{K_1} + (q-1)e^{K_2}}{q}\right]. \tag{6.8}$$

We have now determined both variables in the effective reduced Hamiltonian. Notice that the effective coupling $K'$ depends directly on the WV couplings, $K_1$ and $K_2$, and on the number of values of the $\sigma$ variable, $q$.

## The Effective Ising Coupling

The final step in the partial-summation transformation is to relate the effective reduced Hamiltonian to the Ising model. First, we note that $K'$ in Eq. (6.8) plays the same role as $K_{\mathrm{Liquid}}$ in Eq. (6.1). The only difference is the shift in zero level, which doesn't affect the thermodynamic behavior of the system. It follows that the effective Ising interaction for the WV model can be written as

$$K_{\mathrm{Ising}} = -\frac{1}{2}K' = -\frac{1}{2}\ln\left[\frac{e^{K_1} + (q-1)e^{K_2}}{q}\right]. \tag{6.9}$$

This simple relation connects the WV model to the Ising model.

Next, we introduce the reduced temperature, $T_{\mathrm{r}}$, as was done for the Ising model in the previous section. In particular, we define

$$T_{\mathrm{r}} = \frac{1}{|K_2|} = -\frac{1}{K_2}.$$

This definition recognizes the fact that $K_2$ is negative for all cases of interest in the WV model.

One additional quantity is useful in what follows. Note that for a given mixture of A and B molecules there will be a specific ratio of the bonding to nonbonding energies – a ratio that is characteristic of that particular set of molecules. To specify this ratio, we define the quantity $R$ as

$$R = -K_1/K_2. \tag{6.10}$$

In cases corresponding to closed-loop phase diagrams, $K_1$ will be positive and $K_2$ will be negative, and hence $R$ will be a positive quantity. The larger the value of $R$, the stronger the bonding interaction.

Before we continue, there is one special case that is worthy of note. When $K_1 = K_2 = K_{\text{Ising}}$, which corresponds to $R = -1$, the WV model reduces to the Ising model. To see this, we can write Eq. (6.6) as follows:

$$-\beta H_{\text{WV}} = \sum_{\langle i,j \rangle} \left[ K_{\text{Ising}} \left(1 - \delta_{s_i s_j}\right) \delta_{\sigma_i \sigma_j} + K_{\text{Ising}} \left(1 - \delta_{s_i s_j}\right) \left(1 - \delta_{\sigma_i \sigma_j}\right) \right].$$

Rewriting this equation, and collecting terms, we find

$$\begin{aligned} -\beta H_{\text{WV}} &= \sum_{\langle i,j \rangle} K_{\text{Ising}} \left(1 - \delta_{s_i s_j}\right) \left[ \delta_{\sigma_i \sigma_j} + \left(1 - \delta_{\sigma_i \sigma_j}\right) \right] \\ &= \sum_{\langle i,j \rangle} K_{\text{Ising}} \left(1 - \delta_{s_i s_j}\right). \end{aligned}$$

Thus, we see that this limit of the WV model yields the Ising model. There is no closed loop in this special case, just a single critical point at the top of a coexistence dome, as in Figure 6.17.

## Comparing with Experiments

To compare with experiments, we start by rewriting Eq. (6.9) in terms of $T_r$ and $R$. The result is

$$K_{\text{Ising}} = -\frac{1}{2} \ln \left[ \frac{e^{R/T_r} + (q-1) e^{-1/T_r}}{q} \right]. \tag{6.11}$$

This is the result we will use in all that follows.

We plot $K_{\text{Ising}}$ versus $T_r$ in Figure 6.21 for a variety of values of $R$ and with $q = 8000$. Recall that the critical value of $K_{\text{Ising}}$ for the simple cubic model is $K_c = 0.22165\ldots$ This value is indicated in the figure by the

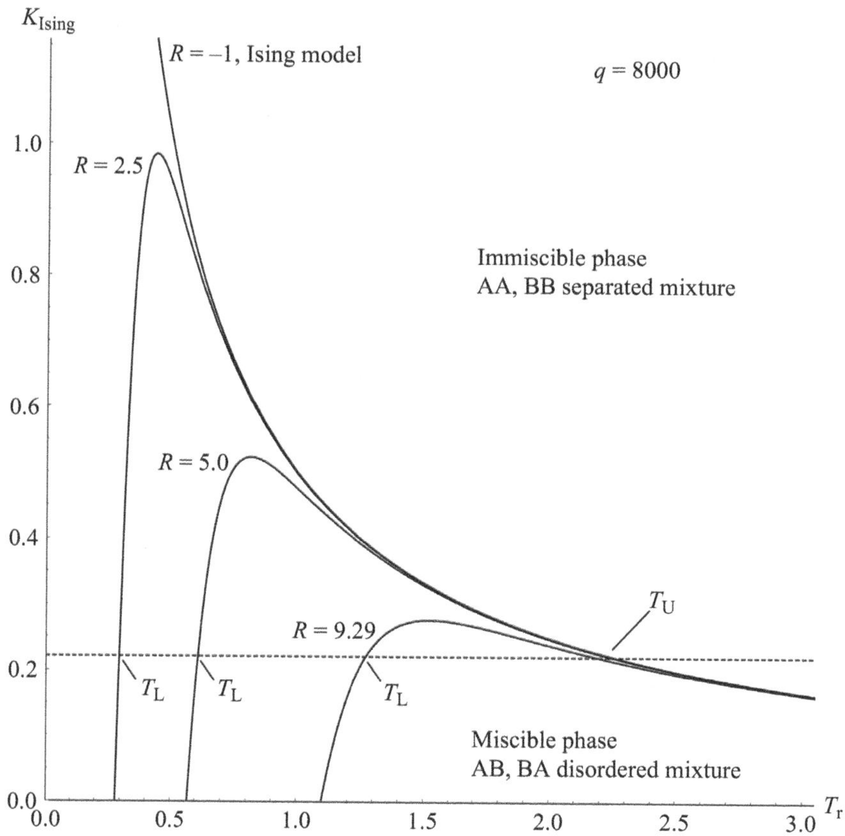


Figure 6.21 The effective Ising interaction, $K_{\text{Ising}}$, versus the reduced temperature, $T_r$, for $q = 8000$. The horizontal dashed line is the critical value, $K_c = 0.22165\ldots$, separating the ordered from the disordered phase of the Ising model. Positive values of $R$ (Eq. (6.10)) yield closed loops between $T_U$ and $T_L$.

horizontal dashed line. When $K_{\text{Ising}}$ is less than $K_c$, the system is disordered; that is, the system is miscible, with A and B molecules mixed together. When $K_{\text{Ising}}$ is greater than $K_c$ the system is ordered; that is, the system is in the immiscible phase, with A and B molecules separated from one another.

The top curve in Figure 6.21 corresponds to $R = -1$, which is the Ising model, as in Figure 6.12. We see that the Ising model has a phase transition at roughly $T_r = 2.2$, indicated by $T_U$ in the figure, where the $R = -1$ curve crosses the dashed line.

The lowest curve in the figure is for $R = 9.29$, which corresponds to a strong bonding interaction. Note that this system is close to the Ising model at high temperature, as one would expect, and it crosses the critical dashed line at the upper critical point of a closed loop, at $T_U \sim 2.2$. As the temperature is lowered further, the curve deviates significantly from the Ising model and dips downward to values of $K_{\text{Ising}}$ that are less than the critical coupling. This results in a lower critical point at the bottom of a closed loop, at $T_L \sim 1.3$.

Thus, at high temperature the hydrogen bonding has little impact. At lower temperatures the bonding is significant, and brings the A and B molecules together in the miscible phase, this time with hydrogen bonds between the components.

The middle two curves in Figure 6.21, for $R = 5.0$ and 2.5, correspond to lower values of $R$, and hence weaker bonding. In these cases, we see that $K_{\text{Ising}}$ tracks with the Ising curve over a larger range of temperatures. The temperature must be lowered further in these cases before the bonding begins to impact the system. Thus, the lower critical temperature for these cases, $T_L$, is progressively lower as $R$ is lowered. This means that a smaller value of $R$ results in a larger closed loop.

The width of a closed loop – that is, the range of composition from one side of the loop to the other – is determined by $q$. We can see this in Figure 6.22, where we plot $K_{\text{Ising}}$ for $R = 2.5$ and a variety of values of $q$. The curve for $q = 8000$ is the same as the curve for $R = 2.5$ in Figure 6.21. As the value of $q$ is reduced, to $q = 1000$ and 100, the height of the curves is reduced as well. A lower height means that $K_{\text{Ising}}$ is closer to its critical values, and hence the magnetization is closer to zero. Recalling that $x = \frac{1}{2}(1 + m)$, a smaller magnetization, $m$, corresponds to a narrower loop, that is, a smaller range of $x$. Thus, increasing or decreasing $q$ increases or decreases the width of a loop.

To understand this $q$ dependence, we note that when $q$ is larger, the entropy loss in going to the bonding level is greater, and hence the bonding doesn't become important until the temperature is lowered further. Thus, as $q$ goes up, the degree of order – that is the value of $K_{\text{Ising}}$ – increases, as we see in Figure 6.22. This causes a wider closed loop, and a smaller temperature for the lower critical point.

We turn now to a specific experimental system. Specifically, the dots in Figure 6.23 are from experiments on glycerol + benzylethylamine. The vertical axis is the temperature reduced by the upper critical temperature. The closed loop corresponds to $R = 9.29$ and $q = 8000$, as in the lowest curve

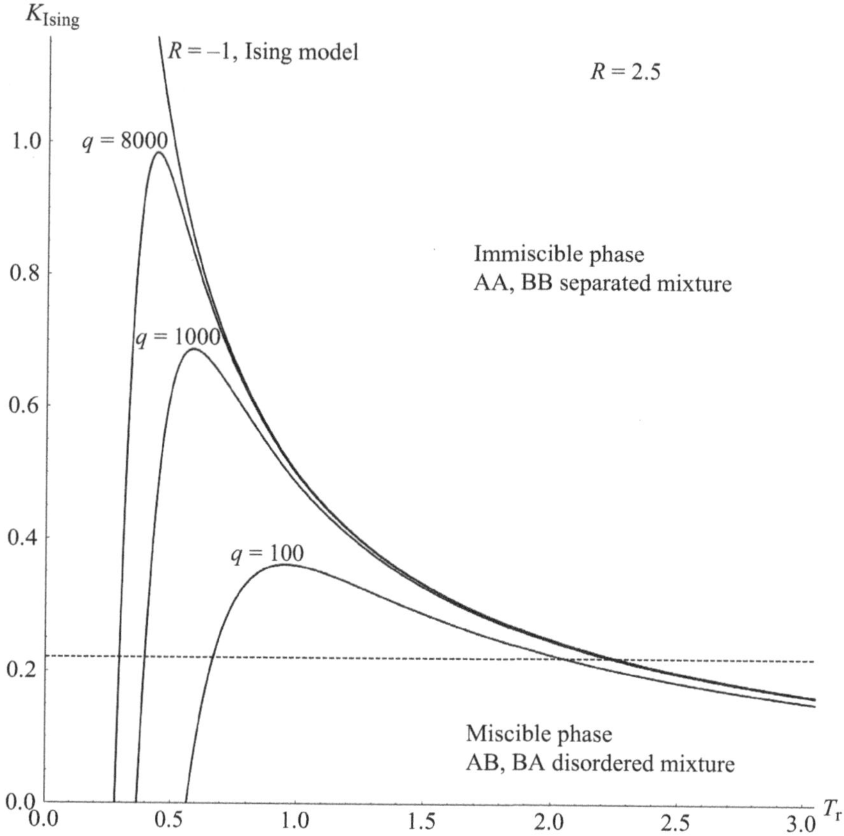


Figure 6.22 The effective Ising interaction, $K_{\text{Ising}}$, versus the reduced temperature, $T_r$, for $R = 2.5$. The horizontal dashed line is the critical value, $K_c = 0.22165\ldots$, separating the ordered from the disordered phase of the Ising model. Notice that as $q$ is decreased the closed loops become shorter and thinner.

in Figure 6.21. Choosing this value of $R$ gives the closed-loop curve in Figure 6.23, with a height going from $T/T_U = 1$ to $T/T_U = 0.58$. The value of $q$ was chosen to give the appropriate width of the loop. We find that $q = 8000$ gives good agreement with experiment. The loop is calculated using $x = \frac{1}{2}(1 + m)$, where for the magnetization $m$ we use the Scesney result given in Section 6.3.

Thus, the WV model has produced good qualitative agreement with experiment with its energy/entropy mechanism. In addition, connecting the WV model to an effective Ising model also allows for quantitative agreement with experiment.

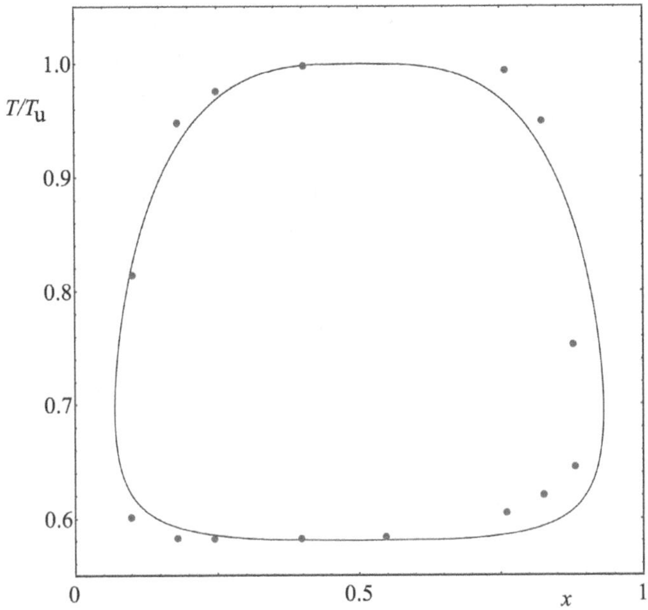


Figure 6.23 Phase diagram for glycerol + benzylethylamine. The dots give experimental results, and the solid curve is the closed loop calculated using $m_{scesney}$ with $R = 9.29$ and $q = 8000$ in the WV model.

## 6.5 Problems

6.1 **Fractional distillation.** Consider a binary liquid mixture of A and B molecules, where the temperature of vaporization for pure A molecules is 100 °C and the temperature of vaporization for pure B molecules is 70 °C. When fractional distillation is applied to this mixture, is the result a liquid that is enriched in A molecules or enriched in B molecules?

6.2 **Two-solid phase diagrams.** Consider the phase diagram in Figure 6.8. Which of the numbered points in the diagram correspond to **(a)** a single phase or **(b)** coexisting phases?

6.3 **Low-temperature expansion.** Consider the lowest-energy excitation from the ground state in an Ising model, as shown in Figure 6.13 for a two-dimensional square lattice. This excitation consists of a single flipped spin. How many bonds are broken for this excitation on **(a)** a three-dimensional simple cubic lattice, **(b)** a two-dimensional triangular lattice, and **(c)** a two-dimensional hexagonal lattice?

6.4 **Low-temperature expansion.** Consider the second-lowest energy excitation from the ground state in an Ising model, as shown in Figure 6.15 for a two-dimensional square lattice. This excitation consists of two flipped nearest-neighbor spins. How many bonds are broken for this excitation on **(a)** a three-dimensional simple cubic lattice, **(b)** a two-dimensional triangular lattice, and **(c)** a two-dimensional hexagonal lattice?

6.5 **Effective Ising interaction.** Consider the WV model with $q = 8000$ and $R = 10$. Find the reduced temperature of **(a)** the upper and **(b)** the lower critical point.

**(c)** If $R$ is increased, does the upper critical temperature increase, decrease, or stay the same? Explain your reasoning.

**(d)** If $R$ is increased, does the lower critical temperature increase, decrease, or stay the same? Explain your reasoning.

**(e)** If $R$ is increased, does the closed loop get larger, smaller, or stay the same? Explain your reasoning.

6.6 **Effective Ising interaction.** Refer to Figure 6.22. Note that $R = 2.5$ in this case.

**(a)** What is the effective Ising interaction, $K_{\text{Ising}}$, when the reduced temperature is $T_r = 0.65$ and $q = 1000$?

**(b)** If $q$ is increased, will $K_{\text{Ising}}$ increase, decrease, or stay the same? Explain your reasoning.

6.7 **Effective Ising interaction.** Refer to Figure 6.21. Note that $q = 8000$ in this case.

**(a)** What is the approximate reduced temperature corresponding to the lower critical point for $R = 2.5$?

**(b)** What is the approximate reduced temperature corresponding to the lower critical point for $R = 5.0$?

**(c)** Is the closed loop corresponding to $R = 2.5$ wider from left to right on a $T$-versus-$x$ plot, narrower, or the same width as the closed loop for $R = 5.0$? Explain your reasoning.

6.8 **Ising limit. (a)** Use Eq. (6.9) to show that the WV model with $K_2 = K_1$ is equivalent to the Ising model.

**(b)** Use Eq. (6.11) to show that the WV model with $R = -1$ also corresponds to the Ising model.

# 7
# Modern Phases and Phase Transitions

The phases we've studied so far – gas, liquid, solid, and magnetic – were known to the ancients. For millennia, no additional phases were known, or even suspected. We now consider some new phases and phase transitions that have only been discovered in the modern era, from the late 1800s to the present.

## 7.1 Plasmas

A *plasma*, from the Greek for "moldable substance," is often considered to be the *fourth state of matter*. Simply put, a plasma is similar to a gas, but one that is characterized by a significant fraction of its particles having an electric charge in the form of positive ions, negative ions, and electrons. As such, plasmas are the most abundant form of visible matter in the universe, from being the main constituent of stars, to dominating the interstellar and intergalactic mediums.

Plasmas were first identified in the laboratory by the physicist Sir William Crookes, who reported on what he called "radiant matter" to the British Association for the Advancement of Science in 1879. Systematic studies of plasmas were undertaken in the 1920s by the Nobel Prize-winning American physicist Irving Langmuir and others. Langmuir coined the term "plasma" to describe what he saw as an analogy between the way an electrical plasma transports electrons and the way blood plasma transports red and white blood cells. Since that time, our understanding of plasmas has increased and deepened significantly.

Though in many respects a plasma can be considered as a low-density ionized gas, there is much more to the plasma phase than that. The plasma and the gas are similar in that neither assumes a definite shape or volume. They are quite distinct, however, in the respect that the particles in a gas have little interaction with one another, while the charged particles in a plasma interact strongly with others, even over a long range. The interaction between particles is not only electrostatic, as one would expect for ions, but as the ions flow in the plasma they create electric currents, which in turn produce magnetic fields, and these combined electromagnetic fields affect the motion of other particles throughout the plasma. This results in a complex, collective, and long-range behavior that is completely absent in a neutral gas.

Let's look at a few common examples of plasmas in the real world.

## Lightning

Though people have always been aware of lightning, its connection to electricity was only proven, by Benjamin Franklin and others, in the mid 1700s. Even then, the details of lightning, and its connection to plasmas, was not known until well over a century later.

As we understand lightning today, the conditions for it to occur begin when turbulent winds cause particles of ice and water to swirl about and collide with one another within clouds. These collisions cause friction between the particles that, in turn, causes charge separation to occur. Positively charged particles concentrate at the top of the clouds, and negatively charged particles dominate at the bottom of the clouds or on the ground.

This separation of charges produces a strong electric field, and when the field exceeds roughly 3 million volts per meter, the molecules in the air are torn apart to form ions in a process known as dielectric breakdown. The result of this ionization is the conversion of the neutral gas molecules in the atmosphere into a conductive plasma. The lightning bolt now transfers massive amounts of charge through the plasma in a fraction of a second, resulting in incredibly high temperatures in the range of 30,000 K – five times the temperature of the surface of the sun. The extreme heat produces the white light of the lightning bolt, and also causes the air to expand rapidly and then to quickly cool again, resulting in the strong CLAP of thunder.

The strongest lightning in the solar system occurs on the planet Jupiter, where it can be 10 times more powerful than terrestrial lightning. It also tends to have a greenish cast due to the high concentration of ammonia in the Jovian atmosphere.

## Aurora Borealis and Aurora Australis

The shimmering northern and southern lights on Earth, aurora borealis and aurora australis, respectively, are due to the interaction of a plasma – the solar wind – with the upper atmosphere at altitudes of 80 km or higher. The solar wind is a diffuse hot plasma of free electrons and positive ions given off from the sun by its outermost layer, the million-degree kelvin solar corona. It reaches Earth with a speed of about 400 km/s, where it interacts with Earth's magnetic field, which is most intense near the poles of the planet.

The energetic charged particles in the solar wind strike molecules in the atmosphere. This excites the molecules to higher-energy states, from which they decay back to the ground state with the emission of light. The resulting green and red colors are due primarily to light emitted by atomic oxygen. Other colors are seen as well, especially blue and purple due to emissions from molecular nitrogen.

## Fluorescent Lights/Neon Lights/Plasma Globe

In 1857, Heinrich Geissler introduced a new form of man-made illumination: the fluorescent light. In these lights, a voltage of 200 to 600 V is applied to a glass tube containing a mixture of argon and mercury. This produces a hot plasma within the tube. When free electrons in the plasma strike mercury atoms they cause the atoms to emit ultraviolet light at 254 and 185 nm. We can't see this light, so there is another essential feature of the fluorescent lightbulb – the inner wall of the glass tube is coated with phosphorous. The purpose of the phosphorous is to absorb the UV light from the mercury atoms and re-emit it by the process of fluorescence as visible light. Thus, these lights are much more involved in their two-step operation than one might imagine.

A similar, but simpler, type of illumination is provided by the neon light. These devices are similar to fluorescent lights in that they consist of a glass tube containing a gas that is ionized by high voltage, in this case about 15,000 volts, to produce a plasma. The difference is that the excited atoms in this plasma emit light in the visible spectrum, so no phosphorous coating is required. For example, neon gas gives off a reddish-orange light, argon gives off blue light, krypton gives off green light, and helium gives off pink light. By combining different elements, and using a variety of different tubes, "neon lights" can produce brilliant glowing displays that can be both festive and eye catching. Georges Claude demonstrated the first neon signs publicly in 1910 in Paris, and within a decade or so these types of lights became quite popular.

Another everyday example of a plasma is the so-called plasma globe. This device, invented by Nikola Tesla in 1894, is a clear glass globe containing a mixture of neon, krypton, and xenon gases. When a voltage of about 5000 V is applied between the inner terminal and the outer globe, a plasma forms, with glowing filaments moving throughout the interior of the globe. Touching the globe can affect the filaments due to the capacitance of the human body allowing a small current to flow through that path.

### Fusion Energy

Another example of a man-made plasma is in devices designed to produce nuclear fusion. These machines, known as tokamaks, use magnetic fields to confine a plasma containing deuterium and tritium within a torus-shaped vacuum chamber. The plasma is heated to temperatures on the order of millions of degrees Celsius, several times hotter than the core of the sun. At these temperatures the deuterium and tritium can fuse to produce a helium nucleus, a neutron, and a release of 17.6 MeV, about four times the energy released in uranium fission. Though nuclear fusion is an attractive method of power generation, using tokamaks or other methods, it has yet to be perfected on a commercial scale.

## 7.2 Liquid Crystals

We turn now to another phase of matter that was not known until late in the nineteenth century. In 1888, the Austrian botanist Friedrich Reinitzer was studying the behavior of cholesteryl benzoate, a chemical derived from plants. As he heated the solid material, he found that it melted at 145 °C, producing a cloudy fluid. On further heating, this milky fluid underwent a second phase transition at 178.5 °C, resulting in a clear liquid. Later that year, the German physicist Otto Lehmann concluded that the cloudy fluid was actually a new phase of matter, and that it had properties that were intermediate between the liquid and crystal phases. He coined the term *liquid crystal* to describe this new phase. Because of its intermediate properties, it is sometimes referred to as a *mesophase*.

The key feature of cholesteryl benzoate, and other liquid crystals, is that it has rod-shaped (prolate) molecules, as indicated in Figure 7.1. In the crystal phase, shown in Figure 7.1(a), the centers of the molecules sit on the sites of a three-dimensional lattice, giving the system long-range positional order. In addition, the orientation of the rods is the same throughout the crystal, giving the system long-range orientational order as well. This is a highly ordered *solid phase*.

At intermediate temperatures, as in Figure 7.1(b), the system is in the cloudy fluid phase – known today as the *nematic liquid crystal phase*. This phase has long-range order in terms of the orientations of the molecules, as in the crystal

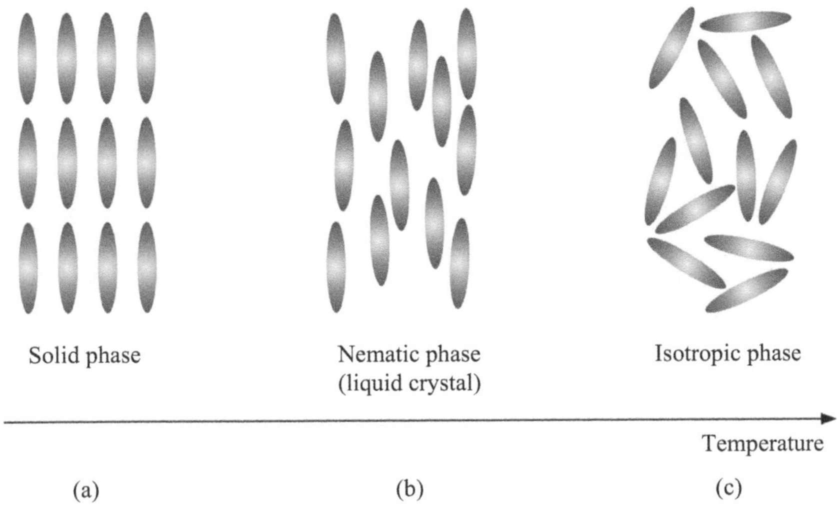


Figure 7.1 Basic phases of a liquid crystal with rod-shaped molecules. (a) The solid phase, where the molecules are aligned in the same direction and arranged in a three-dimensional lattice. (b) The nematic phase, where the molecules are oriented in a preferred direction, but the centers of the molecules are disordered, as in a liquid. (c) The isotropic phase, where the molecules have random positions and random orientations.

phase, but it is disordered in terms of the centers of the molecules, as in the liquid phase.

The term "nematic" comes from the Greek root word nema, which means thread. This is the same root as in the name nematode, which is a thread-like, parasitic roundworm. In the liquid crystal case, nematic refers to the elongated shape of the constituent molecules, and to their alignment in thread-like strips throughout the liquid. To recap, the nematic liquid crystal phase has long-range orientational order, but only short-range positional order.

In contrast, the high-temperature *isotropic phase*, shown in Figure 7.1(c), is disordered both in terms of the positions of the centers of the molecules, and their orientations. This corresponds to the clear liquid phase that Reinitzer saw at high temperature.

There are many other liquid crystal phases that are observed in nature as well. For example, in Figure 7.2 we show two different phases known as *smectic liquid crystals*. The name "smectic" comes from the Greek root meaning soapy, which refers to the visual appearance of these phases. On a molecular level, the key characteristic of smectic phases is that the molecules are *arranged in layers*, stacked one on top of another. Within a given layer,

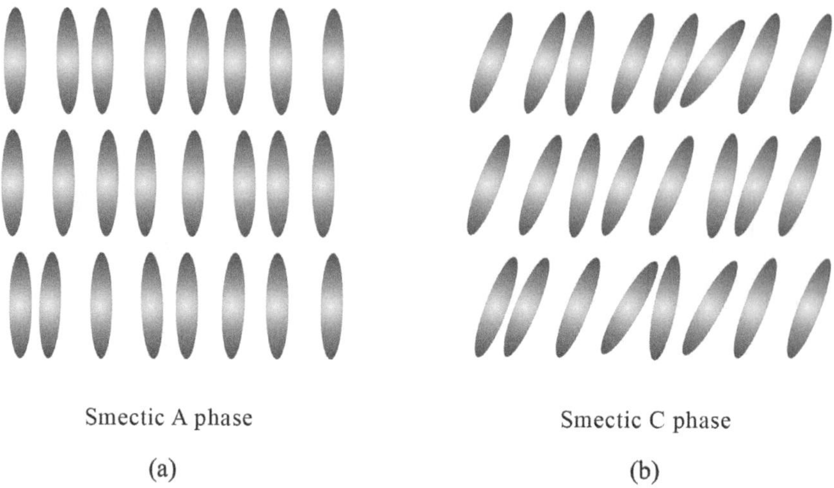


Figure 7.2 Smectic liquid crystal phases, in which the molecules are arranged in layers. (a) In the smectic A phase, the molecules are aligned perpendicular to the layers. Their centers are arranged randomly within a layer, like a two-dimensional liquid. In the smectic B phase, not pictured, the molecules in each layer are arranged in a hexagonal array. (b) The smectic C phase is similar to the smectic A phase, except the molecules are tilted at an angle perpendicular to the layers.

however, the molecules are generally disordered and liquid-like in their distribution.

In the smectic A phase, shown in Figure 7.2(a), the molecules are aligned perpendicular to the layers. The smectic B phase, not pictured, is the same as the smectic A phase except that within the layers the molecules form a hexagonal array. Finally, in Figure 7.2(b), we see the smectic C phase. This phase is again similar to the smectic A phase, except that now the molecules are tilted with respect to the perpendicular to the layers. Within each layer the molecules are again disordered, as in a liquid.

## Liquid Crystal Displays

Many liquid crystal molecules are not only long and thin, but are also *chiral* in nature; that is, they have a right-handed or left-handed helical structure, like that of a helical staircase. As a result, they tend to rotate the direction of polarization of linearly polarized light. This property has become useful in the construction of display screens for computers and other devices.

To be specific, liquid crystal displays are constructed in the following way. First, take a pair of crossed polarizers, which by themselves would not allow light to pass through. Now, between these two polarizers insert a liquid crystal that rotates the plane of polarization by 90°, so that light can now pass through both polarizers. This gives a region of the display that is bright. If, instead, an electric field is applied to the liquid crystal, it causes the molecules to "unwind" sufficiently that the plane of polarization is no longer rotated. This means we're back to the original case with crossed polarizers, and no light passing through – this is, a dark region on the display. By sending electric fields to the appropriate pixels of the display, the liquid crystal can produce a useful and effective display.

## 7.3 Superconductivity, Superfluidity, and Bose–Einstein Condensates

We conclude our survey of modern phases with the discovery of superconductivity and superfluidity in the early 1900s. These phases are inherently quantum mechanical in nature, and as such were completely unanticipated before the advent of quantum theory. Even so, a full theoretical understanding of these "super phases" wasn't achieved until the mid 1900s.

### The Resistivity of Mercury

The discovery of superconductivity occurred in 1911 when the Dutch physicist Kamerlingh Onnes, the first to attain temperatures low enough to liquefy helium, conducted a variety of experiments at ultralow temperatures. In one series of experiments, Onnes measured the electrical resistivity of various metals – first solid mercury, and later tin and lead – as a function of temperature.

Onnes's experiments were exploring a realm of temperatures lower than had ever been studied before. As a result, no one knew exactly what to expect. William Thomson (Lord Kelvin), a master of classical physics, believed that as the temperature was lowered electrons flowing through a metal would slow and eventually come to a complete halt – corresponding to infinite resistivity. Onnes, on the other hand, knew from previous experiments that resistivity tended to decrease with temperature. Thus, he expected the resistivity to decrease steadily to zero as the temperature went to zero. It turns out that neither of these predictions was correct.

What actually happened is shown in Figure 7.3. Here we show the resistivity, $R$, of a mercury wire as a function of temperature, $T$, in the neighborhood of 4.0 K. Initially, the resistivity drops roughly linearly as the temperature is

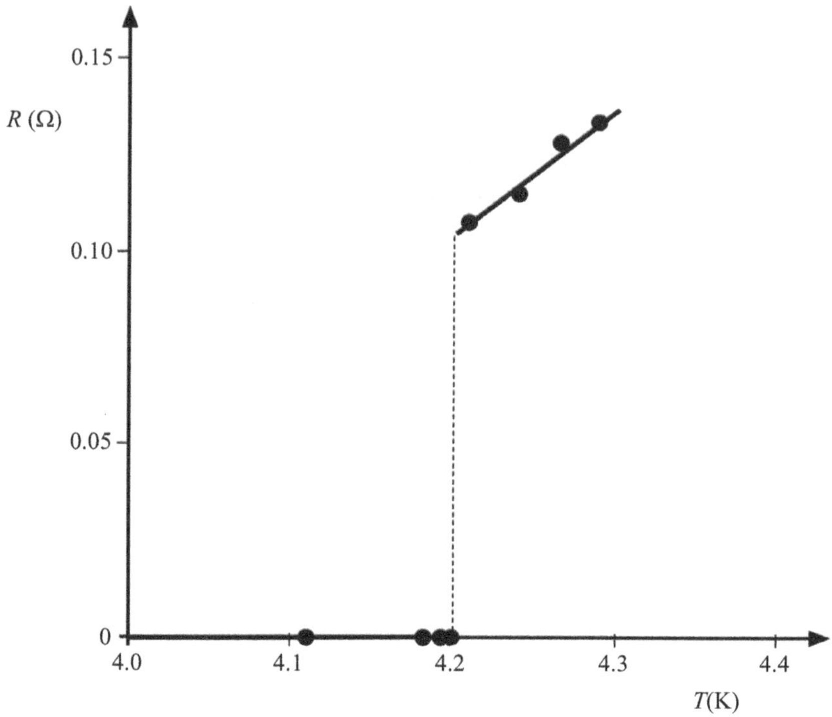


Figure 7.3 Schematic representation of the resistivity of mercury near 4.0 K. Notice the drop to zero at 4.2 K, indicating the onset of superconductivity.

decreased, as expected by Onnes. But then, at about 4.2 K, the resistivity drops suddenly to *zero* – and stays there for lower temperatures.

Onnes recognized the significance of these results and wrote the following in his notebook:

> Mercury has passed into a new state, which on account of its extraordinary electrical properties may be called the superconductive state.

This was, indeed, the first observation of superconductivity. Similar transitions to superconductivity were observed by Onnes in tin (at 3.7 K) and lead (at 7.2 K); this was apparently a widespread phenomenon.

## Some Properties of Superconductors

There are many striking properties that set superconductors apart from other phases of matter. For one, the fact that the resistivity of a superconductor is zero means that if an electric current is established in a superconductor it can persist forever. For example, if an electric current is established in ring of superconducting material, the current can continue undiminished indefinitely, as long as the temperature is maintained low enough to accommodate the superconducting state. This has been demonstrated experimentally, showing that the current can persist for at least 100,000 years.

Another landmark property of superconductors is the ability to exclude a magnetic field from their interior. For example, if a magnetic field is turned on near an ordinary conductor, the conductor responds by inducing electric currents that oppose the change in the magnetic field. This is the Lenz effect, and is an example of diamagnetism. Superconductors, with their zero-resistance current flow, are perfect diamagnets, and hence they completely oppose the increasing magnetic field and exclude the field completely. This is referred to as the *Meissner effect*, after experiments by Walther Meissner and Robert Ochsenfeld in 1933. The Meissner effect persists only up to a certain maximum magnetic field, at which point the superconducting phase breaks down and converts back to the normal phase.

A particularly striking demonstration of the Meissner effect is provided by magnetic levitation in high-temperature superconductors. These superconductors are referred to as high-temperature because their transition temperature from superconducting to normal is greater than 77 K, the temperature of liquid nitrogen. Since liquid nitrogen is easy to obtain, as compared to liquid helium, which is needed for superconductors like mercury or tin, these high-temperature superconductors are useful for practical applications, such as MRI machines.

In magnetic levitation a disk of a high-temperature superconductor is placed in a bath of liquid nitrogen, and a small magnet is placed on the disk. When the temperature of the disk drops below the critical temperature the Meissner effect kicks in and the permanent magnet is levitated. This is a striking visual manifestation of superconductivity.

These high-temperature superconductors are produced by compounds of various ceramics. The first to be described, and the best known to this day, is yttrium barium copper oxide, with a critical temperature of 93 K. These new types of superconductors exploded onto the scientific scene in 1987, and resulted in a legendary marathon session at that year's annual meeting of the American Physical Society in New York. This "electrifying" session has been referred to as the "Woodstock of Physics." Great excitement surrounded this development, which continues to this day because the promise is so high. In addition, the mechanism of this type of superconductivity is still not fully understood theoretically, and hence remains an area of active research.

The superconductivity of pure elements, such as that of Onnes's mercury, is referred to as type I superconductivity. In this type of superconductivity, the transition from superconducting to normal as the magnetic field is increased is sharp and abrupt. In high-temperature ceramic superconductors, the transition from super to normal with increasing magnetic field is gradual, mediated by magnetic vortices that grow in number until the super phase is destroyed. These superconductors are referred to as type II.

## The Mechanism of Superconductivity

To understand how superconductivity works on a microscopic level, we start with a discussion of resistivity in a normal metal. The basic idea of electrical resistivity is that as electrons – the classical charge carriers of electric current – flow through a crystal lattice, they interact with and scatter from the ions that form the lattice. This is somewhat like a person (the electron) trying to move through a crowded room full of people (the ions). As the temperature is lowered, the ions oscillate about their crystal locations with smaller amplitude, and hence are less effective in scattering the electrons. Thus, the resistivity of a metal tends to decrease with decreasing temperature, as seen on the right-hand side of Figure 7.3.

To understand how the resistivity drops to zero at 4.2 K, we must go beyond classical physics and include quantum mechanical effects. One of the basic aspects of quantum particles is that particles with half-integer spins, like electrons and protons, are fermions, and cannot occupy the same quantum state. On the other hand, particles with even spins, like photons and phonons, are bosons, and readily occupy the same state. An example are the photons in a laser that form a coherent beam.

The basic idea of superconductivity, then, is that if the charge carriers are bosons, they can condense into the same quantum ground state. In addition, if the ground state is lower in energy than the first excited state, so that there is an energy gap between these states, there will be no excitations if the thermal energy is less than the gap energy. This means that at low temperatures the charge carriers can flow in the ground state, suffering no scattering, and hence with no resistivity. The only problem with this picture is that electrons – the charge carriers in the classical case – are fermions, not bosons. We address this issue next.

## Cooper Pairs and BCS Theory

How can a physical situation consisting of electrons flowing through a metal be equivalent in some way to a system of bosonic charge carriers? This connection was first established by the American physicist Leon Cooper in

1956. His insight was that electrons in a metal can form *bound pairs*, today known as *Cooper pairs*, that behave as bosons because they contain an even number of fermions. These pairs can then condense in a quantum ground state to form a superconducting phase.

The question now becomes, how can electrons – which in a vacuum repel one another due to their like charges – feel an *attractive interaction* in a metal and form a bound pair? To see how this can happen, imagine an electron flowing through a crystal lattice of positive ions. As the electron moves along, it attracts positive ions in its neighborhood. The ions move toward the electron, but because of their inertia in this dynamic situation, they overshoot the position where they would perfectly screen the negative charge of the electron, and hence they produce a net positive charge around it instead. This positive charge can then attract another electron, with opposite spin and momentum, and form a bond between them.

This is sometimes referred to as the *mattress effect*, where one electron pulls in positive ions, like a person depresses a mattress, and another electron is attracted to it, like a second person feels an attraction toward the first person's depression in a mattress. The Cooper-pair mechanism is also responsible for the *isotope effect* in superconductors. If a metal is made with a different isotope, so that its positive ions have a different mass, this changes the response of the ions to the negative charge of the electrons. For example, if a more massive isotope is used, the positive ions respond less and produce a smaller excess positive charge around the electron. This results in weaker bonding of the Cooper pair, and hence a lower transition temperature for superconductivity in that isotope.

With the Cooper-pair mechanism now in place, the stage was set for a comprehensive theory of superconductivity based on the idea of a phase transition involving condensation of bosonic Cooper pairs. Such a theory was produced by the American physicists John Bardeen, Leon Cooper, and John Robert Schrieffer in 1957, and is known today as BCS theory. They were awarded the Nobel Prize in Physics for this work in 1972.

BCS theory is a form of mean-field theory, in a quantum context, and in many respects is similar to the mean-field theory of magnetism presented in Chapter 5. As such, BCS theory predicts that the order parameter for superconductivity, which is essentially the density of Cooper pairs in the condensed phase, vanishes near the critical point as the square root of the reduced temperature, as is characteristic of mean-field theories. Deviations from mean-field theory, as expected in a more complete theory that includes the effects of fluctuations, only become important very close to the critical point. For almost all other temperatures, the mean-field description works well.

## Superfluidity and Bose–Einstein Condensates

Superfluidity is a phenomenon that has many close connections to superconductivity. Both occur at cryogenic temperatures, and both are fundamentally quantum in nature, characterized by the condensation of bosons into a "super" phase. In a superconductor, the flow of electric current occurs without friction or resistance; in a *superfluid*, the liquid itself flows without resistance.

It's possible that Kamerlingh Onnes produced a superfluid in his experiments with liquid $^4$He in 1911, where he discovered superconductivity. In any case, extensive experiments into superfluidity in $^4$He were conducted by the Soviet physicist Pyotr Kapitsa, and independently by the Canadian physicists John F. Allen and A. D. Misener, in 1938. In $^4$He, a system in which the individual molecules are bosons, the condensation into a superfluid phase occurs at about 2.17 K.

Just like superconductors, superfluids have extraordinary properties that are not seen in "normal" fluids. As a superfluid, $^4$He flows without resistance through even the smallest pores in any container. For example, if normal $^4$He is held in a container without leaks, it's possible that leaks will appear once the liquid has become super. In addition, a superfluid in a container will climb up the walls in a super version of capillary action, and eventually flow out of the container, emptying it even if there is no "leak." Superfluids are also incredible conductors of heat, with a heat conductivity that is millions of times greater than in the corresponding normal liquid, and even greater than in the best metallic conductors.

As mentioned, the individual particles in $^4$He are bosons, given that each $^4$He atom consists of an even number of fermions. As a result, it's reasonable to expect liquid $^4$He to have a superfluid phase transition. On the other hand, $^3$He is a fermion, and hence its molecules are not capable of undergoing a bosonic condensation. However, just as electrons can bond together as bosons in Cooper pairs, $^3$He atoms can form pairs as well, producing *quasiparticles* that have bosonic properties. These quasiparticles can then condense into a superfluid.

Another quantum phase that is an area of active current research is provided by what is referred to as a *Bose–Einstein condensate*, referred to as a BEC. In such systems, a cloud of bosonic molecules is held by magnetic confinement and reduced in temperature with laser cooling to incredibly low temperatures, typically on the order of $10^{-7}$ K. This is a whole new temperature range, with new types of physical behavior. In 1995, the American physicists Eric Cornell and Carl Weiman at the University of Colorado, along with Wolfgang Ketterle at the Massachusetts Institute of Technology, succeeded in producing a BEC, and were subsequently awarded the Nobel Prize in Physics in 2001.

One of the many extraordinary characteristics of a BEC is that it consists of a shared quantum state on a macroscopic scale. This system has many striking properties, such as an incredible slowing of light. For example, a BEC can slow the speed of light within it from $3 \times 10^8$ m/s to just 17 m/s in a typical case. In addition, in some experiments it has been possible to stop a pulse of light altogether within a BEC, and to release it later.

Clearly, a great deal of promise exists in studying and understanding the properties of the modern phases that have been discovered in the last couple of centuries. Some of these are fundamentally quantum mechanical in nature, others simply depend on unusual properties of the elementary constituents of the substance. New physics, and new behaviors and applications, await researchers as additional phases of matter will surely continue to be discovered in the future.

## 7.4 Problems

7.1 **Liquid crystal.** If a substance is in the nematic phase, what phase do you expect to occur when the temperature is raised? Explain your reasoning.

7.2 **Liquid crystal.** A given substance can be in the nematic phase or the smectic phase. Is the temperature of the smectic phase greater than, less than, or the same as the temperature of the nematic phase? Explain your reasoning.

7.3 **Superconductivity.** The phase transition to superconductivity in mercury occurs at roughly 4.2 K. If a lighter isotope of mercury is used instead, will the phase transition to superconductivity occur at a temperature that is greater than, less than, or equal to 4.2 K? Explain your reasoning.

# Index

Printed in the United Kingdom by TJ Clays Ltd.